Why Stuff Falls Down

The Story of Science and Gravity From Aristotle to Einstein Told in Plain English

By Michael Kemp

Revisions made in 2025: Fixed typos, spelling, and grammatical errors and rewrote a handful of sentences & paragraphs for increased clarity.

Library of Congress Control Number: 2024905387

This book is intended for curious high schoolers, advanced middle schoolers, homeschoolers, unschoolers, and life-long-learners of all ages.

SCI033000	SCIENCE / Physics / Gravity
YAN050120	YOUNG ADULT NONFICTION / Science & Nature / Physics
EDU017000	EDUCATION / Home Schooling

Published by:
What a Curious World LLC
WhatACuriousWorld.com

Table of Contents

"The most exciting phrase to hear in science,
the one that heralds new discoveries,
is not 'Eureka!' but 'That's funny...'"
~ Isaac Asimov

Table of Contents

Acknowledgments

The parts of this book that are easiest to understand? That's thanks to my long-suffering wife Danita Reynolds and her insistence on clarity. Thanks also go out to Eric Roost, my sons, and other family and friends for their enthusiasm and encouragement. Geoff Keyes caught a couple of goofs I made early on, much appreciated.

My gratitude to Emma Lyddon, for her videography on the chapter activity videos, and her enthusiasm and encouragement.

Many thanks to Ray Morse and Paul Thornton for their editorial skills, encouragement, and probing questions. Any grammatical or punctuation errors that you find were undoubtedly made (by me) *after* Ray or Paul had any chance to correct me.

Sincere appreciation goes to Brian Bray for his encouragement, and particularly for his suggestion of adding activities to the book!

I grew up near the Oregon Museum of Science and Industry. I'm sending out an æthereal "Thank you!" to my late parents for the many visits we made to OMSI. My science curiosity continued to be fed by an endless stream of sci-fi books and *Star Trek* episodes. And by the TV shows *Cosmos* and *The Day the Universe Changed*. A big shout-out goes to the 1966 book by I.S. Shklovskii & Carl Sagan *Intelligent Life in the Universe.* Popular science books just don't get any better than that.

I am deeply indebted to Mr. Hawken, who taught some college-level physics to us bonehead high school students and kindled my lifelong love of physics.

If, like me, you are hungry for science curiosities and current developments, here's a sampling of YouTube channels to feed your inner nerd:

- *PBS Space Time* by Matt O'Dowd — a deeper look into all things space and time.
- *Star Talk* by Neil deGrasse Tyson — fun, jocular science, astrophysics, and more.
- *Arvin Ash*, *MinutePhysics*, and *The Science Asylum* are (all three) nerdy but great channels, with explainers on a wide variety of physics questions.
- *Fermilab* — particle physics presented in (mostly) plain language.
- *PBS Terra* — news and curiosities mainly dealing with critters and biology.
- *Cool Worlds* — intriguing podcasts about cosmology, exoplanets, and more.

But seriously, as far as acknowledgments go, I gotta give it up for this amazing, mind-boggling, vast, and truly bizarre universe — and the tiny speck of it that we inhabit. And for the curiosity, zaniness, and occasional genius of us humans.

We are a curious breed!

A Note on Using the Reference Links

There are way over 100 reference links in this book. They are mostly noted by a small number hovering over the text, like this.[0] If you look at the end of each chapter, you will see that number next to a web link and a QR code, like this:

Deeper Dive
0. What a Curious World website
https://www.whatacuriousworld.com/

How to use these links?

- If you are using a digital version of the book: just click on the text link.

- If you are using a printed version of the book, with a phone or tablet:
    - Point your camera app at the square QR code
    - When the camera app puts brackets around the QR code you can click on the QR code itself or on the button saying "Open in Browser"
        - If you do not see brackets around the QR code, go to Settings/Camera and check that QR Code recognition is "On."

- If you are using a printed version of the book, with a computer:
    - In your web browser go to WhatACuriousWorld.com
    - In the menu bar at the top of that web page hover over "The Book!"
    - A drop-down menu will appear
    - Select the chapter that you are working on
    - On that chapter's page, click on the "Jump to Links" button (or just scroll down until you see the links)
    - Click on the desired link

What do the link colors mean?

Blue for Biography: Find out more about this person.
Green for Deeper Dive: Go here for more in-depth information about a topic.
Purple for Great Video: I've scoured the web looking for the most straight-forward and awesome videos to illuminate the subjects at hand.
Red for Physics-Speak: Wowzers! Is this how they really communicate? Why, yes it is. And they have really good reasons for using the bizarre-but-specific language of science and mathematics.

Links to websites and videos were valid when I made them. The author (that's me) cannot be responsible for changes made to other folks' websites. And, of course, advertisements on said websites can be obnoxious — even disgusting — but again, not within my control. Let's hope for the best.

1. We Got This, Right?

"One thing I have learned in a long life:
that all our science, measured against reality, is primitive and childlike
— and yet it is the most precious thing we have."
~ Albert Einstein

What's This Book About?

There are many things that tickle my old brain cells. One of the most curious and irritating subjects to itch the gray matter is that thing we glibly call "gravity." I mean, how complicated can it be? Stuff falls down. Something as straightforward as stuff falling down should be totally figured out. Straightforward facts, straightforward explanation, surely.

It turns out that a lot of folks have tried to explain gravity. From ancient times up to this very day, philosophers and physicists have come up with one theory after another to explain gravity.

Now, I do a lot of name-dropping in this little book but don't worry about keeping all of the characters sorted out. I give out names for a couple of reasons — first, to give credit where credit is due, and second, so that if you are curious, you can research this or that historical person on your own.

From ancient to modern times, one theory of gravity has replaced another, each one more accurate than the ones that came before. But, even as the older theories have been disproved, each and every one of these theories is useful in its own way.

So if the earliest theory of gravity was useful, then why did folks bother to create all the later theories of gravity?

Because, my inquisitive reader, while an earlier theory might be useful in its limited way, there always seemed to be some situation in which the theory would fall apart. Some real-world facts would not fit into what the old theory said should happen. So what then? The challenge was to construct a new theory of gravity, which would not only cover all of the things that the old theory got right but also hold together for the situation(s) where the old theory fell apart.

Facts, Hypotheses, Theories (and Gravity)

Before we get started we need to clear up the difference between what a **fact** is, what a **hypothesis** is, and what a **theory** is.

Facts are raw data. Every time I hold a bowling ball chest high and just let it go, it falls down and smacks the floor. And it smacks the bowling alley's pretty maple slats at my feet with the same force each time I drop the ball from chest height. That is, until the manager throws me out. So, it is a fact that when I let go of a bowling ball at chest height, it falls downwards and smacks the floor.

"Hypothesis" is a fancy word for an educated guess. A hypothesis is some idea we humans come up with to explain the facts. It takes in the facts about real-world phenomena (like falling bowling balls) and sets out a system of ideas to explain how or why those facts occur. Any hypothesis worth its salt has to make some unique predictions that can be tested. A hypothesis is a proposed theory that has yet to be tested.

A **theory**, on the other hand, is a hypothesis that *has* been tested and supported. If any of the predictions that a hypothesis makes fail a real-world test, it doesn't make the cut. A theory is a successful set of ideas that explain how or why the facts occur. A theory does the explaining. A theory is how we humans strive to understand the facts that nature puts in front of us. The facts just keep rolling along, whether our current theory about them is rock solid, or rickety as an old 3-legged fruit picker's ladder.

Warning: In this little book, I am using a fairly strict "scientific method" definition of the word "theory." In daily conversation, and even in science literature, the word "theory" is often used to mean a guess. In daily conversation, "theory" often means what should really be called a "hypothesis." But in serious science, you don't get to call your wild idea a theory unless it makes unique predictions that are supported by experimental results.

A physicist acquaintance objected to my using the word "fact." He noted that using the word "fact" implies that the thing being talked about cannot be questioned. For instance, common sense tells me that, when I'm sitting on a park bench, I am stationary. And, since I don't feel any movement, the Earth itself must be stationary. Sitting on my park bench and watching the Sun cross the sky, common sense tells me that the Sun is circling the Earth. It would be easy to think that the Sun circling the Earth is a *fact*.

A scientist would prefer to say that watching the Sun cross the sky is an "observation" or "observed phenomena." Any ideas or conclusions that you draw from this observation are not facts, they are just ideas, and if they hold up to experiment, they might even become accepted theories.

So, I can see the physicist's objection to using the word "fact" because, on the one hand it is easy to confuse your own interpretation of an event as a fact. And on the other hand, the word "fact" implies the thing being talked about is fixed, that it is something that cannot be questioned.

My physicist acquaintance makes a valid argument against using the word "fact." But, I'm going to keep using "fact" interchangeably with "observation," "observed phenomena," "experimental results," and so forth. I just ask you to keep in mind that I'm using the word "facts" for stuff that actually happens in the world, not what we *think* we know about it.

And by the way, it took a lot of folks a long time to sort out that even though it *looks* like the Sun circles the Earth — that's not what's really going on. Check out this link for a deeper dive into the history of how we figured out the structure of our solar system![1]

In science, all observations, even the most obvious, are open to questioning. Science is all about observing, developing ideas about what has been observed, and testing those ideas against the real world.

Unquestioned facts, common sense, and The Truth are not what science is about. Science is about how the universe *actually* works. Not how we *want* it to work.

And **gravity**? Gravity is the term we use when we are talking about how, or why, stuff is pulled downwards. Gravity is not the bowling ball falling down. Gravity is the explanation for how or why a bowling ball falls down when you let go of it. The bowling ball dropping to the floor is a fact. Gravity is a theory.

Over the millennia, folks have come up with quite a few different theories about gravity. Each theory gets tested against reality and that's when the theory shows just how solid, or rickety, it really is. As you will see shortly, it sometimes takes centuries before anybody bothers to do any serious testing. But hey, each theory was made by a celebrated genius and only a crazy person wants to go up against a celebrated genius, right?

Come along with me, and let's take a look at what these "geniuses" said about why stuff falls down.

What Direction Is Your Down?

What sort of a dork doesn't know where "down" is? Well, me for one. Unless you tell me *where* you are, I have no idea what your direction for down is!

When you think about it, gravity is just our word for "stuff falls down." And what does "down" even mean? When I trip and fall and hit the ground things like "gravity" and "down" are very real to me.

If gravity is just our word for "stuff falls down," then let's look at what we mean by "down." It's not as straightforward as common sense tells us it should be.

If someone lives in Hawaii, way out in the Pacific Ocean, then "down" for them is the same direction as "up" is for somebody who lives in Botswana, way over in Africa, because the two of them are on opposite sides of this big round rock that we call Earth. The *direction* of "down" for our friend standing in Hawaii points from their head to their feet and through the center of the Earth and out the other side and straight through Botswana and out into space. If our friend in Botswana looked in that direction (out into space) well, our friend in Botswana would say that direction is *not* "down" it's "up!" So, is the *direction* of "down" in Hawaii really down? For someone in Hawaii, why, yes it is. For folks anyplace else on the Earth, not so much. Especially if they live in Botswana.

Like many things in physics, **the direction of down varies depending on where the observer stands.** That doesn't mean that you can have any direction you choose to be "down" — it just means that the direction of "down" *here* is different from the direction of "down" *over there.*

We could even say that our reality (the direction of down in this case) changes depending on our location.

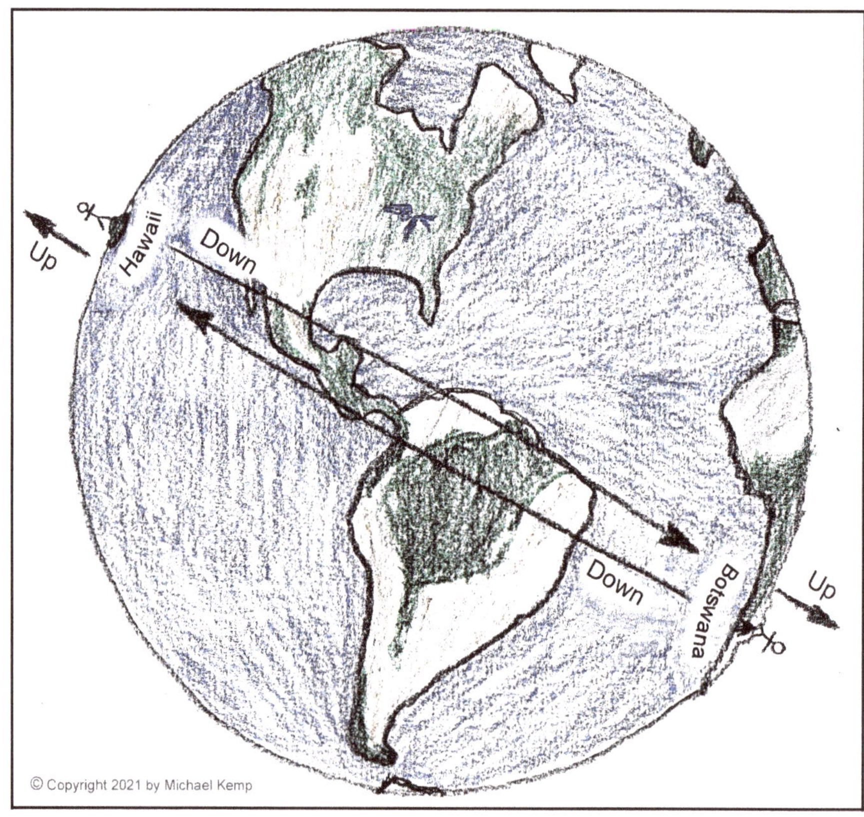

Fair warning: more of our reality is relative to our location than we'd like to think about. It's not *just* the direction of down. We will return to this point later on, with a vengeance.

At the human scale of existence, these variations by location are like a strange flutter at the corner of our vision; slightly unsettling, but of no real consequence in our day-to-day activities. We just tell ourselves, "I know what down is" and get on with our daily lives.

Stuff

As you may have gathered, this chapter is setting the stage for our exploration of why stuff falls down. We've just talked about "down," but what is "stuff?" When I say "stuff" I mean anything you can hold, breathe, drink; anything made up of any form of matter. To get technical, I'm talking about anything made of atoms or elementary particles.

The idea that stuff (most of which can be seen with the naked eye) is made up of atoms (which cannot be seen) is quite a leap. Why would you think that stuff that you *can* see is made from immense numbers of atoms that you *cannot* see?

Most folks don't spend their time thinking about what makes up stuff, why it falls down, why the Sun moves across the sky, why there are seasons, how lightning happens, or any of that. Not today, not a century ago, not thousands of years ago. But the folks who *did* think about such things in ancient times decided that the gods must be to blame.

The Sun was a god named Ra by the ancient Egyptians, or Inti by the South American Incas, or Helios by the ancient Greeks (they had Helios driving a golden chariot across the sky; that's one shiny chariot). And the same sort of explanations were concocted for other natural phenomena. If anybody happened to think about something so constant and so mundane as stuff, and why it falls down, well then, the gods just *made* the world that way. End of story, and everybody was happy. Well, almost everybody.

Over the millennia there have been a few thinkers around the planet who have come up with ideas about natural phenomena that didn't rely on saying something like, "The gods did it." But most of the time folks were more comfortable believing that the gods did it, so there was no need to try to understand what was going on.

One exception to this cropped up in ancient Greece. Many Greek philosophers had thought that stuff was a manifestation of the abstract, archetypal substances of earth, water, fire, air, and æther.

In case you were wondering, æther (the ancient Greek archetypal substance) is often spelled ether. It is pronounced "eethr" — which sounds just like the other word ether. The other ether is a flammable chemical used as starter fluid and in anesthesiology. I'm using the spelling "æther" so that you'll know that I'm talking about the archetypal substance, not the starter fluid. Æther was believed to be an all-pervading medium that permeated all of space.

Most ancient Greek philosophers believed that these archetypal substances were uniform and continuous. It was believed that real-world stuff could manifest in whatever quantity was required, and each piece of stuff existed as a solid block. You could cut a piece of stuff into smaller and smaller pieces. The only limit to how small you could divide up your stuff was the limitation of your tools. It was believed that there was no lower limit to how small you could cut up stuff.

Then things got messy. A little over two millennia ago some Greek guys named Leucippus,[2] Democritus,[3] and Epicurus[4] put forward the idea that, nah, stuff isn't completely solid and continuous, it's made up of huge numbers of tiny, indivisible, uncuttable particles, suspended in the void. They called these uncuttable particles "atomos" from the prefix "a" (not) and the Greek word "temnein" (to cut). In modern English, we've shortened that to **atom**.

Leucippus, Democritus, and Epicurus did not have the atom smashers and cloud chambers that we have these days to observe atoms and subatomic particles, but their atomic theory — that stuff is made out of conglomerations of tiny uncuttable particles that are so small that they cannot be seen — was amazingly close to what has been experimentally verified. With the exception that the tiny bits of stuff that came to be called atoms turn out not to be the smallest of particles, and what we today call "atoms" are not *completely* uncuttable.

Elements and Atoms and Particles, Oh My!

What we now call an "atom" is not truly uncuttable, and when you throw in words like "elements" and "particles" it can get a little confusing. It all has to do with how human understanding of stuff (matter) changed over the centuries.

As folks organized our knowledge about different types of stuff it became clear that some stuff (an apple, for instance) was made up of a mixture of simpler substances, while other stuff (gold, for instance) was a pure substance all by itself. These pure substances were obviously the elemental building blocks of everything that we think of as "stuff." It became accepted that these elementary substances (hydrogen, helium, oxygen, carbon, gold, etc.) were each made up of their own type of atom. For instance, one type of atom was an atom of gold, and if you could scrape together a humongous number of these gold atoms it would make a lump of gold that you could use to make beautiful jewelry — or trade for food, shelter, clothing, tools, or other goodies.

So each pure elemental substance (hydrogen, helium, oxygen, carbon, gold, etc.) was made of its own distinct type of atom. And calling each pure substance an "element" stuck. As a result, we talk about "The Periodic Table of the Elements," which is really just a list of all of the types of atoms that we have so far detected.

But atoms are not actually *uncuttable*. The elemental substances listed in The Periodic Table of the Elements are not actually the *most elementary* building blocks of matter, precisely *because* these atoms can be busted up into smaller bits.

Atoms have been broken down into **particles** called protons, neutrons, and electrons. Protons and neutrons, in turn, have been broken down into particles called quarks. It doesn't look like quarks (or electrons) can be broken down any further, so in physics-speak, quarks and electrons are called **elementary particles**.

To summarize, in modern physics-speak:

- An **element** is a pure substance, like oxygen, carbon, gold, etc.
- An **atom** is the smallest possible building block that still has the properties of that pure substance.
- A **particle** is any tiny bit of matter or force. It can be an elementary particle, like an electron, or some particles (like protons and neutrons) are built out of even smaller particles (quarks, in this case).
- **Elementary particles** are the *really truly* smallest bits of uncuttable stuff in our universe. They can be a particle of matter (like an electron) or a particle of force (like a photon).

So all of these names for bits of stuff got a little mashed up over the course of time, but once you see how the names got assigned, it makes a little more sense.

But Is It Real?

So this theory about atoms is cool, and all that, but are atoms real? When I say that atomic theory has been experimentally verified, that's putting it mildly. These days the manufacture of everything from synthetic materials to computer chips is based on our current theories about the physics and chemistry of atoms. Medical procedures from X-rays to MRIs are based on theories about how atoms are constructed and how they interact with electromagnetic rays and pulses.

And it's hard to ignore that the theories about how atoms can be built up or broken apart allowed us to figure out how to build nuclear weapons — both how to break atoms apart as in the fission bomb (that would be "atom bomb classic," like the ones that destroyed Hiroshima and Nagasaki) and how to mash little atoms together to form bigger atoms as in the fusion bomb (the new and improved "hydrogen bomb" which has been tested in remote places and underground but, thankfully, has not been used on any cities just yet).

The theory of atoms in general, and the theories about how atoms operate, have certainly been put to work in ways that are sometimes subtle and sometimes not so subtle. The theories work.

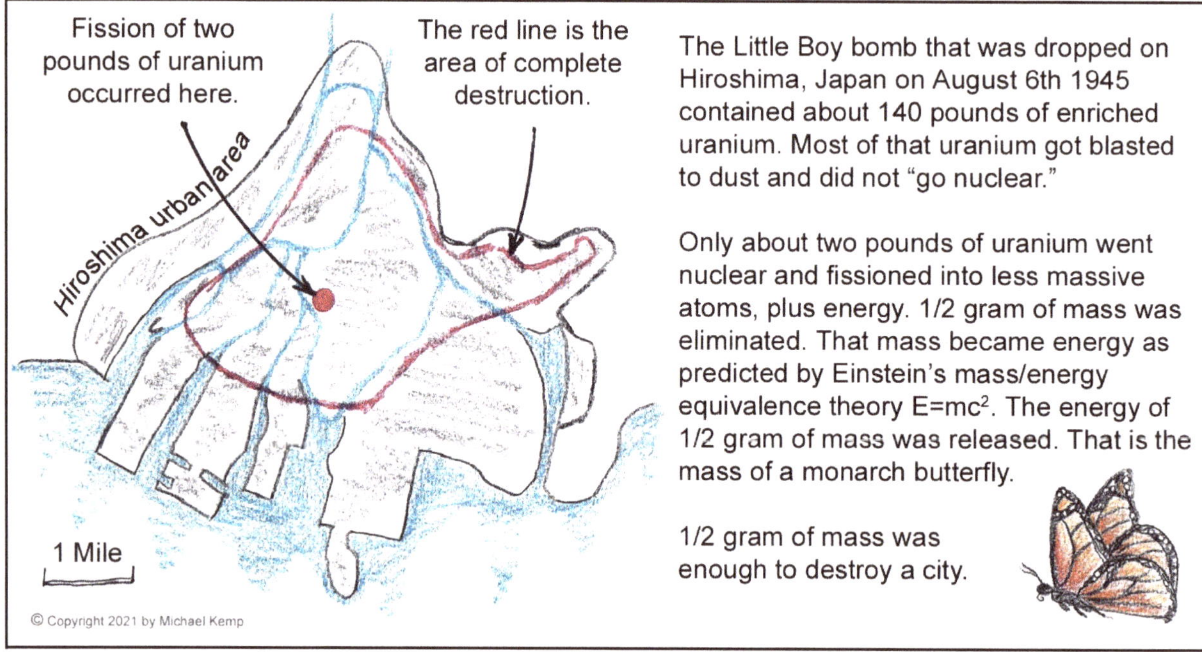

As an aside, that Greek guy I mentioned, Democritus, was an interesting character. He inherited a fair amount of wealth and spent a bunch of it traveling in foreign lands. He looked up teachers and philosophers and magi. When he returned home, he delved into **natural philosophy** — which is a fancy way of saying that he studied how things work. Plus he became quite a writer and proponent of both the theory of atoms and of **scientific rationalism**. Scientific rationalism is a fancy way of saying, "Let's look at what is right in front of us and not try to cram what we observe into something that Zeus decreed." You could also call this approach **empiricism** or **empirical research**. He may have had his taste for scientific rationalism fortified by his visits to Egypt and Babylon, where empiricism had roots going back at least another millennium.

Democritus seems to have been amused by life, the universe, and everything. Folks called him "The Laughing Philosopher," but that's another story.

Leucippus, Democritus, Epicurus, and other ancient Greek philosophers were doing their best to see the world rationally, and to think critically about reality — rather than how their myths and beliefs told them it should be. This caused conflict with the established norms of their society. So it got messy.

The messy part of it was that while the priests railed against scientific rationalism and critical thinking as attacks on their ancestral beliefs and piety in general, they were unable to stop the freight train of rational and critical thought. Rational thinking and critical thinking are two of the foundation stones of the scientific method. Without them (and the principles of repeatability, prediction, and peer review) we'd still be saying that things fall down because some god commands it.

The priests and protectors of propriety in ancient Greece did occasionally win the day, as when they convinced 280 out of the 500 Athenians in the jury to condemn Socrates to death[5] for impiety and for corrupting the youth of Athens. Both of these charges were well-founded and he was, indeed, guilty of them. He was impious by not properly honoring the traditional gods revered by Athenians. He corrupted the youth of Athens by shaking their unquestioning faith in their city's government, gods, and priests. He taught the kids a form of critical thinking that we now call the Socratic method.[6]

As much as any of the ancients, Socrates laid a solid foundation for science as we know it. His aphorism "The only true wisdom is in knowing you know nothing" embodies the ruthless questioning and continuous testing that sharpens the edge of scientific progress.

Meanwhile In India…

Greek philosophers were not the only ones to dream about atoms. I would be remiss, in discussing atoms and stuff, if I did not mention the ancient atomist teachings of India. Some Indian philosophers embraced the idea of anu (which was their name for atoms).

Six centuries BCE, the Vedic philosopher Kanada put forth a detailed description of the atoms that he believed were the basic building blocks of our world.[7] Similar to the ancient Greek philosophers, Kanada described the elementary nature of matter as earth, water, fire, air, and æther. In addition, Kanada includes the elements of space, direction, time, mind, and soul in his enumeration of the anu (atoms) from which existence is constructed.

More than a millennium later, in the seventh century CE, a Buddhist philosopher named Dharmakirti again brought atomism into the Indian limelight.[8]

All of this is to say that the concept that *all the stuff that we experience is built from atoms* had been put forward by various ancient philosophers from various traditions.

As noted above, in some Indian traditions, anu (atoms) are the building blocks not just of matter, but of direction, space, time, and consciousness itself.

Curiously enough, there are echoes of this view of space and time in current efforts to reconcile Einstein's general relativity with quantum mechanics.

There are also contemporary musings that consciousness may be built into the fabric of our universe (as put forth in the ancient Indian idea of anu of consciousness). Other folks think that consciousness is an emergent property of the nervous system, or a gift from the gods, or an expression of being-ness from some other dimension. But concepts of consciousness are far beyond the scope of this little book.

Moving right along, in the next chapter we will return to ancient Greece, and a guy named Aristotle who was creating a lot of the groundwork for European philosophy and science. He even came up with a theory about how stuff falls down. It did not rely on the will of the gods.

What a mess.

Review

Take a moment to think about what makes a fact a ***fact*** (or if you prefer — what makes an observed phenomenon an ***observed phenomenon***)? What makes a ***hypothesis***? And what makes a ***theory***?

The word "atom" is derived from ancient Greek and means uncuttable. What we have called **atoms**, as the centuries rolled by, turned out to be tiny structures that *can* be cut up into smaller bits. Big atoms can be split apart into smaller atoms. This is what powers our current fission-based nuclear power plants, and how the WWII-era atom bombs worked. Multiple atoms can also be merged into bigger atoms. The fusion of hydrogen into helium powers our Sun using this process. This same fusion process powers modern nuclear bombs.

But an atom still qualifies as the smallest bit that you can have of a pure element (like oxygen, carbon, or gold). And yet, atoms themselves are constructed out of even tinier bits of matter: electrons, protons, and neutrons. Protons and neutrons, in their turn, are constructed out of quarks. As far as we can currently tell, that's as deep as the rabbit hole goes.

"Why stuff falls down" sounds simple, but the closer you look at "stuff" and at what "down" means, the curiouser and curiouser it gets.

Activity

Our Round Planet, and the Direction of Down

Here's a slightly goofy activity, but it helps get a feel for how our big round rock called the Earth has a million directions for "down." This sets the stage for thinking about our planet, stuff, and gravity.

For this activity, you'll need:

- One round (or round-ish) balloon. If you have a ball that you can mark up, you can use that instead of the balloon, if you wish.
- A printout of a map of the Earth that shows all of the continents. I'm rather fond of the Goode projection (Google "Goode projection" with the "e" in Goode).
- A marking pen to mark up the balloon. If you've got small stickers around, those are handy to mark special places (like where you live).
- A table lamp or floor lamp.

If any of the following is confusing, please watch the video of me doing this activity in the link below. Besides, in that video I do a couple of extra demonstrations with the toy Earth to show how day-any-night and the seasons work (that's what the table lamp or floor lamp is for).

Blow up the balloon. The balloon represents the Earth. The knot that you tied to seal the balloon? Let's call that the South Pole. Take the marker and put an X on the opposite end of the balloon. That's the North Pole. Draw a dashed line around the middle of the balloon, half way between the North and South poles. That is your Earth balloon's equator.

Use your printed map of the Earth to draw in the continents. They don't need to be detailed, or even all that accurate — just so that you can say "There's Africa, there's North and South America, there's Europe and Asia, there's Australia." Mark an X on the balloon showing where you live, or even better, mark it with a sticker!

Keep in mind that, from any location on Earth, "down" points to the center of the Earth. OK. Now, which direction is "down" at the point on your Earth where your home is? Which direction is "down" from each of the continents? "Down" truly depends on location, location, location.

Watch the video at this link to view this activity ***plus*** demonstrations of why we have day-and-night and seasons:
https://www.whatacuriousworld.com/chapter-1/#Video01

Links

Deeper Dive
1. "Planetary Motion: The History of an Idea That Launched the ...." 7 Jul. 2009, https://earthobservatory.nasa.gov/features/OrbitsHistory.

Biography
2. "Leucippus - Wikipedia." https://en.wikipedia.org/wiki/Leucippus.

Biography
3. "Democritus - Wikipedia." https://en.wikipedia.org/wiki/Democritus.

Biography
4. "Epicurus - Wikipedia." https://en.wikipedia.org/wiki/Epicurus.

Deeper Dive
5. "Trial of Socrates - Wikipedia." https://en.wikipedia.org/wiki/Trial_of_Socrates.

Deeper Dive
6. "Socratic method - Wikipedia." https://en.wikipedia.org/wiki/Socratic_method.

Deeper Dive
7. "Acharya Kanada: The Father of Atomic Theory - Sanskriti Magazine." https://www.sanskritimagazine.com/vedic_science/acharya-kanada-father-atomic-theory/.

Deeper Dive
8. "Dharmakirti - Wikipedia." https://en.wikipedia.org/wiki/Dharmakirti.

2. Common Sense Ain't Always Right

"Many fail to grasp what they have seen,
and cannot judge what they have learned,
although they tell themselves they know."
~ Heraclitus

In looking into how folks have thought about gravity, I like to start with the ancients and work forward from there. The earliest theory of gravity that we have from the ancient Greeks may seem a bit naïve, but it has a practical quality that is hard to deny.

Aristotle Pontificates

Aristotle[9] was born in 384 BCE. Democritus was still alive and in his 70s, and while The Laughing Philosopher avoided the limelight, Aristotle thrived on public appearances and shone brightly on the stage of life. In lectures and writing, he pontificated on everything from physics and logic to politics and poetry to psychology and zoology to… oh you get the picture. Aristotle's work is a foundation stone of Western culture and science.

It should come as no surprise that, among his innumerable pronouncements, Aristotle also formulated a theory about stuff that falls down. He explained that stuff was attracted to the center of the universe, which was the Earth, and the heavier the stuff was, the greater the attraction and the faster it would fall. As well as having a thing for an Earth-centered universe, Aristotle threw in a lot of philosophizing about the five basic elements of existence: earth, water, fire, air, and æther. He took it as self-evident that the core of the universe was the Earth, where the heaviest element would reside, so Earth is composed of, well, earth. The next sphere surrounding earth was the next lighter substance: water, then surrounding the sphere of water was the sphere of air, then came the sphere of fire, and finally, the sphere of æther — where the sun and stars reside.

If a bit of something found itself in the wrong sphere, it would seek its rightful place in its proper sphere. Therefore when bubbles of air are released in water, they float up to rejoin the sphere of air. When an object that is made of earth substances, such as a hammer, is released in the air, it falls down to join its proper sphere, earth.

So the takeaway was that everything seeks its own sphere and that heavy stuff falls faster than light stuff. That's pretty obvious, right? A hammer falls faster than a feather. Case closed. This was accepted as The Truth for nigh unto two millennia.

Aristotle's theory was just common sense. But what is common sense, anyway? **Common sense** is developed from the mental shortcuts we learn to get by in day-to-day life. If you have the repeated experience that a hammer falls faster than a feather, it becomes common sense that heavy stuff falls faster than light stuff. Common sense also tells us that the Earth is flat and stationary and that it is the Sun that moves across the sky. Unfortunately, in many cases, the mental shortcuts of common sense lead us to false beliefs.

And yet, sometimes, as the centuries roll by, a curious person will question common sense and perform an experiment that parts the veils of belief and reveals the stranger nature of our universe.

Leonardo da Vinci took a peek through the veils of Aristotle's description of gravity somewhere around the year 1500. In his private notes, disguised in his "mirror writing," are diagrams and calculations that show that he understood that objects accelerate as they fall. This was contrary to Aristotle's assertion that objects fall at a constant speed (which they reach quickly at the beginning of a fall). Leonardo's calculations came within a few percent of the actual rate of acceleration of Earth's gravity. Not bad.

Stevin and de Groot Drop the Ball

You've probably never heard of these guys, but in 1586 a couple of nerds in the Netherlands decided to test how things fall down for real. Simon Stevin and Jan Cornets de Groot (and you thought nobody could honestly say, "I am Groot") climbed 30 feet up in a church and started dropping stuff. Historical accounts are a little vague, but it seems that they performed two separate experiments. In each experiment, two balls were dropped at the same time. In each test, the balls were constructed in such a way that one ball was 10 times heavier than the other. In one test, both balls were made of lead, one smaller than the other. In the other test, both balls were of the same size, but one was lead and the other was of a lighter material. Now according to Aristotle, objects fall at a speed proportional to their weight. So according to Aristotle, the heavier ball should fall at 10 times the speed of the lighter ball, because it was 10 times heavier.

Nieuwe Kerk in Delft, where Stevin & de Groot proved Aristotle was wrong about how stuff falls down.

Without a high-speed camera to show how much time elapsed between the heavy and the light ball hitting the floor, our intrepid experimenters judged the end of their fall based on the sound of the balls striking a wooden platform. In both experiments, the balls were released together, and in both experiments, the two balls could be heard hitting the wooden platform simultaneously.

This was not what Aristotle had said would happen. Aristotle's theory was considered The Truth for 1,800 years. And yet, it was disproved in a matter of minutes by two guys dropping balls in the Netherlands.

Galileo Gets Theoretical

Also around this time, in the late 1500s and early 1600s, far away to the south, in Italy, there was a guy you probably *have* heard of. Galileo Galilei[10] is considered to be the father of modern science. If you hear the term Scientific Revolution, you will most likely hear Galileo's name mentioned too. It was a heady time when many European scientists were openly questioning "The Truth" about how the world was put together.

Careful measurements, repeated experiments, and rational and critical thinking were used to challenge not only the traditional knowledge passed down from the ancient Greeks but even to challenge Church doctrine.

Once again, the priests were not amused. They went so far as to burn one outspoken scientist, Giordano Bruno,[11] at the stake for his insistence that the Earth was not the center of the universe, that there are multiple worlds, and that the Earth circles the Sun. It probably didn't help Bruno that he insisted on being snarky and disrespectful of his inquisitors. But still, being burned alive seems a little harsh.

Even though Galileo expressed many of the same heresies as Bruno, he got off light. Or at least, without being lit up like a torch. The Church convicted him of heresy[12] and put him under house arrest in 1633, where he remained until his death in 1642. Galileo certainly did not help his case when he published a fictional dialogue titled *Dialogues Concerning Two New Sciences* where he named his opponent in these dialogues "Simplicio" (implying that his detractors were simple-minded).[13]

Throughout his life, and even after his conviction for heresy, Galileo pursued experiments, theories, and inventions in diverse areas of physical science. Among his many interests were the "laws of motion." His thoughts about gravity have a lot to do with his experiments with balls rolling down inclined raceways and with timing the swings of pendulums.

It is unclear whether Galileo heard of Stevin and de Groot's experiment in Delft, but just three years later, in 1589, Galileo made a proclamation about dropping balls from the

Leaning Tower of Pisa. It is tempting to speculate that Galileo developed this idea after getting reports of Stevin and de Groot's Delft experiment. In any case, you could say that he picked up the balls and ran with them.

Galileo proposed that if two balls of different weights were to be dropped at the same time from the top of the Leaning Tower of Pisa (Pisa was Galileo's hometown), they would strike the pavement at the same time. There is some academic disagreement as to whether Galileo actually dropped a couple of balls from the tower or whether he just lectured about what would happen *if* somebody were to drop two balls of different weights from the tower. In any case, he generally gets the credit for the type of experiment that Stevin and de Groot actually performed.

But what's more, Galileo thought about it deeply enough to come up with a better theory about how stuff falls down. Since the facts gleaned from Steven and de Groot's experiment knocked the blocks out from under Aristotle's theory of gravity, it was obvious to Galileo that a new theory of gravity was called for.

One problem with Aristotle's theory is that a light object (let's say a feather) and a heavy object (let's say a hammer) interact with the medium that they are falling through (air) in very different ways. The feather catches more air in relation to its weight than the hammer does, so air friction slows the feather down a lot. Air friction hardly slows the hammer down at all. The drop-two-balls experiments minimized that difference in air friction, and in so doing it tested whether weight, by itself, makes a difference in how stuff falls down.

In Galileo's *Dialogues Concerning Two New Sciences*, his personal avatar (Salvaiti) states that: **a heavy body has an inherent tendency to move with a constantly and uniformly accelerated motion toward the common center of gravity, that is, toward the center of our Earth**, then he goes on to say: **a heavier body does not move more rapidly than a lighter one provided both bodies are of the same material.**

Galileo goes on to clarify that in specifying the "same material" he was referring to the role that air friction plays in slowing down a falling object. If air friction is removed, such as in a vacuum, he maintains that gravity accelerates all objects at the same rate. He uses an inflated bladder and a lead weight as an example of objects that would fall at the same rate in a vacuum, but not through the air.

This is the heart of Galileo's theory of gravity. If air friction is removed as a factor, objects of any type will fall at the same rate. And he maintains that objects are continuously accelerated by gravity. The Leaning Tower of Pisa and Steven and de Groot's experiments confirmed that gravity affects all objects equally, regardless of their weight. Galileo did numerous experiments rolling balls down inclined planes and timing the ball's movement through segments of their descent to confirm that gravity continuously accelerates objects.

Galileo's theory of gravity was a huge leap forward from Aristotle's.

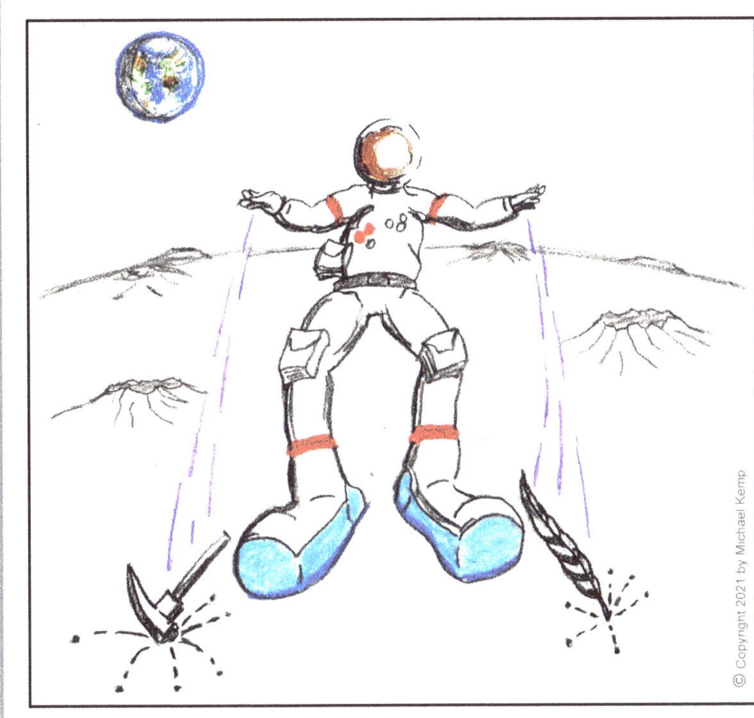

Just for grins and giggles, in 1971 while bounding around on the moon, Apollo 15 astronaut David Scott performed a demonstration of Galileo's theory of gravity. Standing in the vacuum of space, he dropped a hammer and a feather to see if they would fall at the same speed. Follow this link[14] to see the video.

For another test, follow this link[15] to watch a bowling ball and a batch of feathers being dropped from the top of a giant vacuum chamber right here on Earth, in Cleveland, Ohio.

Aristotle's theory was not totally incorrect. It was a rough approximation of everybody's daily experience and explained pretty well how the world works.

Galileo's replacement theory did a good job of showing why Aristotle's theory failed the dropping-lead-balls test, and Galileo's theory did a *better* job of understanding how the world works. Galileo's theory was not perfect either, but it was more useful than Aristotle's.

And that, my patient reader, is how science progresses. A theory holds sway unless and until it is disproved. Then, hopefully, it will be replaced by a new theory that takes into account the latest experimental results. Each refinement of theory brings us to a more accurate understanding of how this universe actually works. How we put that knowledge to use is our responsibility.

Most of the time, common sense is just plain reliable and useful. Don't stick your hand in boiling water. Don't walk off the edge of a cliff. That sort of thing. But "heavy stuff falls faster than light stuff" — it ain't necessarily so.

While science, in one form or another, had been practiced in ancient times, scientific progress had crept through the centuries at a snail's pace. Something happened in

Galileo's time that allowed science to really take off. In the next chapter, we look into the disciplined approach that gave science the wings to fly: **the scientific method**.

Review

The ancient Greeks had a poetic way of dividing the world up into five "elements." From the heaviest to the lightest, those elements were: earth, water, fire, air, and æther. In Aristotle's view of reality, the Earth was the center of the universe, to which everything else was attracted. The heavier an object's elemental nature, the more it is attracted to the center of the Earth. If an object finds itself in an element that it is not composed of, it will seek to return to the sphere of its true elemental nature.

According to Aristotle, if you dropped two balls, one weighing one pound and the other weighing two pounds, the two pound ball would travel twice as fast as the one pound ball. They would each travel at a constant speed, but their speed would depend upon their weight.

When Stevin and de Groot tested this idea for real, it knocked the blocks out from under Aristotle's theory.

When Galileo made a better theory of gravity, he basically said that if you can eliminate air friction as a factor, everything falls the same. And stuff does not fall at a constant speed. It accelerates.

Activity

Let's Drop the Ball!

Let's try Stevin and de Groot's ball drop for ourselves!

For this activity, you'll need:

- One lightweight plastic die from a game.
- A pebble that weighs at least twice as much as the plastic die. Or you could use a heavy metal die, or a "shooter" marble. Basically, any hard, roundish object that is at least twice as heavy as the plastic die.
- A kitchen scale, or a letter scale, would be helpful to weigh the die and pebble — but not required.
- If you do this inside, use a piece of cardboard (or some such) to protect the floor.

If you have a kitchen scale or letter scale, weigh your plastic die and your pebble (or whatever). How many times heavier is the pebble than the die? If the pebble is twice as heavy as the die then Aristotle said it will fall twice as fast as the die. If it's three times as heavy, Aristotle said it will fall three times as fast. And so on.

If you are indoors, place the cardboard on the floor in front of you. If you are outdoors, find a patch of concrete or something that will make a satisfying "bang" when the die and pebble hit it.

Raise the die and pebble as high as you can over the cardboard or concrete. Let them both go at the same time. If the pebble is twice as heavy as the die then, *according to Aristotle*, the die will have fallen only halfway by the time the pebble hits the ground. If the pebble is three times as heavy then, *according to Aristotle*, the die will only be one-third of the way down. If it's four times as heavy, then the die will only be one-quarter of the way down, and so forth.

Who do you think had a more accurate idea about how gravity works, Aristotle or Galileo?

Here's a video of me performing this activity:
https://www.whatacuriousworld.com/chapter-2/#Video02

Links

Biography
9. Aristotle - Wikipedia."
https://en.wikipedia.org/wiki/Aristotle.

Biography
10. "Galileo | Biography, Discoveries, & Facts | Britannica."
https://www.britannica.com/biography/Galileo-Galilei.

Biography
11. "Giordano Bruno | Biography, Death, & Facts ...."
https://www.britannica.com/biography/Giordano-Bruno.

Deeper Dive
12. "Galileo affair - Wikipedia."
https://en.wikipedia.org/wiki/Galileo_affair.

Deeper Dive
13. "Dialogue Concerning the Two Chief World Systems - Wikipedia."
https://en.wikipedia.org/wiki/Dialogue_Concerning_the_Two_Chief_World_Systems.

Great Video
14. "Apollo 15 Hammer-Feather Drop - YouTube." 20 Jul. 2015 https://www.youtube.com/watch?v=oYEgdZ3iEKA.

Great Video
15. "Brian Cox visits the world's biggest vacuum chamber - YouTube." 4 Nov. 2014, https://www.youtube.com/watch?v=JWeNToW9t58.

3. Science Settles On a Method

"I would rather have questions that can't be answered
than answers that can't be questioned."
~ Richard Feynman

The foundation stones for science were laid down by the ancient philosophers of Greece, Egypt, Mesopotamia, and beyond. To my mind, the most important of these philosophical ideas were empiricism, rational thinking, critical thinking, and skepticism.

Empiricism is putting more trust in what you actually encounter in the world, than in whatever it is that you might dream up in your head. The results of physical experiments are to be trusted. Ideas or attitudes that are in conflict with real-world experience are to be questioned.

Rational thinking means that you compare one of your thoughts or beliefs to all of the other thoughts and beliefs that you hold to be true. If there are inconsistencies, you question the thoughts and beliefs that are in conflict. Your actions are also part of this picture. If you act one way and think another way, then either your thought or your action is irrational.

Rational thinking means that you strive to have a worldview that is not in conflict with itself. Your thoughts, beliefs, and actions should be logically consistent.

Critical thinking, on the other hand, means that you hold your own convictions up to the light of day. You actively question whether those convictions are justified by the facts at hand. Rather than just trusting what feels right, or what you have been told is The Truth, you examine whether your convictions are in accord with what actually goes on in the world.

Skepticism requires that you hold off judgment until the facts are in. To be skeptical is to suspend belief in what you are being told, or in what you think about a situation, until there is convincing real-world evidence. Skepticism, and an openness to questioning your own beliefs, are the hallmarks of critical thinking.

Before the general acceptance of rationalism and critical thinking, the priests and pundits could explain away natural phenomena by saying, "The gods want it that way, so don't worry your pretty little head about it."

When rationalism and critical thinking gained credibility in ancient Greece, more folks actually did start to wonder (without resorting to this or that god) how and why things work the way that they do, including thinking about why stuff falls down. But while this got science as we know it started, it was slow going.

To sort out solid scientific theories from wild ideas, the Iraqi mathematician, astronomer, and physicist Hasan ibn al-Haytham (born 965, died 1040) advocated a method of science that relied on a repeating cycle of observation, hypothesis, experimental results, and independent verification. Unfortunately, ibn al-Haytham's methods did not catch on right away. In fact, it took until the late 1500s and a man named Francis Bacon,[16] before it really took hold.

The formalizing and adoption of this framework in the late 1500s, called **the scientific method,** kicked science into high gear. The scientific method created such a leap forward in science, that the next hundred years are referred to as the **Scientific Revolution**.

In any sincere discussion of the Scientific Revolution, you will be certain to hear these two names mentioned: Galileo Galilei, and Sir Francis Bacon. While Galileo was an experimenter at heart, Bacon was an organizer.

Sir Francis Bacon was devoted to libraries, even developing a system for cataloging books. Law and politics were his main career. Sir Francis Bacon rose to Attorney General, personal legal counsel to Queen Elizabeth I, and Lord Chancellor of England under Queen Elizabeth I and King James I. So, yah, Bacon was really into law, order, and organizing stuff.

Bacon was a contemporary of Galileo, and while Galileo labored in Italy to build new scientific theories based on experimental data, Bacon's greatest contribution to science was to hammer out a method to judge whether this or that scientific idea represented how the world *actually* works.

Bacon was the obsessive grownup, going around putting everything away in neatly organized patterns — while Galileo was the perpetual kid-at-heart, always curious, always wondering how things worked and what would happen if…

Bacon's scientific method was grounded in his fondness for empirical evidence. The scientific method declares how one should go from a guess, an inspiration, a conjecture, to the creation of an experimentally supported and respected scientific theory. More importantly, the scientific method weeds out any "theories" that are *not* supported by real-world testing. In the words of Richard Feynman:

> *"The first principle is that you must not fool yourself —*
> *and you are the easiest person to fool."*
> *~ Richard Feynman.*

Here's my $1.00 tour of how the scientific method works (the bullet point version):

- Form a question about some natural event or observation.

- Make some guesses about what the answers to that question might be.

- Compare your guesses against existing knowledge, then pick one guess that matches all the available data. You can call this a **hypothesis**.

- Any hypothesis that can be taken seriously must make unique predictions. Those unique predictions must be so clearly stated that they can be tested in the real world. When a physicist says that "a theory must be falsifiable" that's their fancy way of saying that it must make specific enough predictions that testing can show whether the prediction is true or false.

- If the result of testing of *any* prediction of a hypothesis shows that prediction to be false, then the hypothesis is disproved. At that point, the hypothesis must be discarded or revised.

- If all of the results of such tests agree with the hypothesis's unique predictions it is well on its way to being an actual scientific-method-certified **theory**.

- To share a new theory with the world, the researcher(s) write up a paper describing the theory, its predictions, the tests they have run, the data they used, and how they analyzed that data. This can be posted as a **pre-print** to an online site like Cornell University's arXiv.org.

- To become *really* official, the researcher(s) will submit their paper to a respected and specialized journal such as Science. The journal's editor will forward their paper to two or three physicists working in the same field. This is **peer review**. These reviewers look for originality, logical consistency, clarity, the validity of the test data, and the validity of the methods used to analyze it. If the reviewers tell the editor that the paper passes muster, the journal will print the paper. If not, the paper may be rejected or sent back for revision. It can take as much as a year before the paper will actually be printed. The journal Science only prints about 8% of the papers that it receives.

- Once the paper is published (either in a journal, or as a pre-print), scientists around the world can look at the logic behind the theory and try to replicate the test results of the theory's predictions. The more tests that are performed, and the more respected scientists there are performing those tests, and the more consistent the results of those tests are — then the more convincing those test results are, either in supporting or disproving the theory.

And here's the visual version:

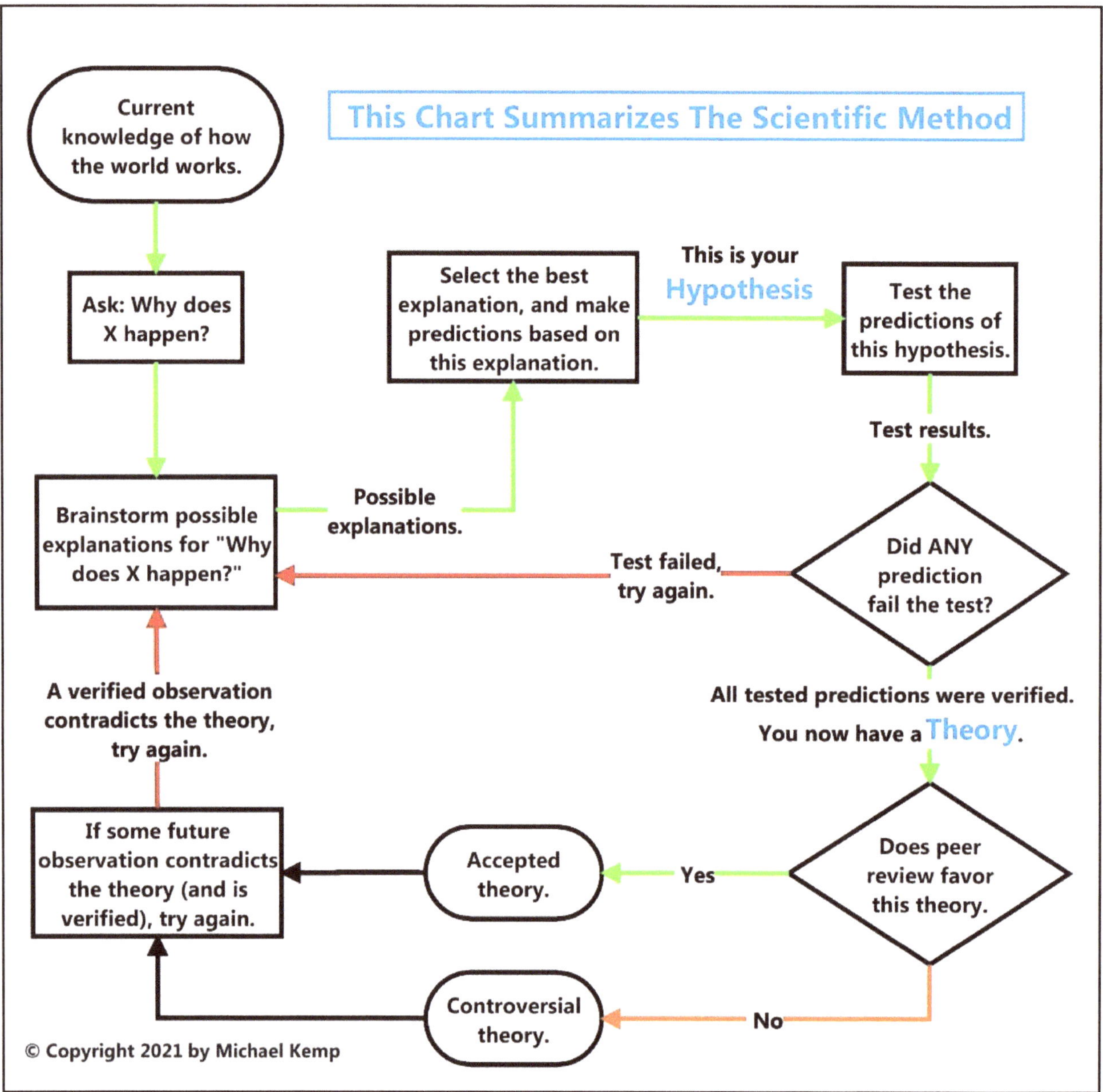

Warning: There is some sloppiness in how the word "theory" gets tossed around, even in scientific literature. Sometimes a hypothesis, whose unique predictions have not yet been tested, is *called* a theory — while *I* think it should still be called a hypothesis.

String theory, I'm looking at *you*!

In science, a theory can be disproven by experimental results that contradict predictions made by that theory. But even when all the experiments support a theory, a scientific theory is never truly "proven." The best you get is that the theory is experimentally supported and widely accepted.

The predictions of a theory must be specific enough to be expressed in mathematical terms, so that they can be tested. This is *how* the theory should work *in practice*. For a scientific-method-certified and supported theory, these are called **Laws of Nature**.

But even what has been taken to be a Law of Nature for centuries may later be disproven. We've just seen how the Laws of Nature in Aristotle's theory about how stuff falls down were disproven by a simple test.

As far as *proving* a theory, there are proofs in mathematics and there are proofs in logic but there are no absolute proofs in science. The best that you can get from the scientific method are theories that are well supported by experimental results and are accepted by peer review.

I'm also here to tell you that "The Truth" is not to be found in science. The scientific method is all about observing and questioning and testing and questioning all over again. The goal of science is to understand how our universe works, not how we *want* it to work.

This is an important distinction, which split science off from philosophy, metaphysics, mysticism, and religion, so I'll reiterate it. Experimental results can support a theory. Those experimental results have to be repeatable by other scientists. Plus it really helps the credibility of a theory when it is subjected to critical review by other scientists in the field of study (aka "peer review"). Even then, a theory is never considered the end of the scientific endeavor. On the other hand, experimental results can *disprove* a theory.

A theory that is supported by experimental results and peer review represents our best understanding at that time of how the universe is put together.

"The public has a distorted view of science because children are taught in school that science is a collection of firmly established truths. In fact, science is not a collection of truths. It is a continuing exploration of mysteries." ~ Freeman Dyson[17]

The scientific method gave science the solid footing that it needed to move forward faster. The exponential increase in the speed of scientific progress, that followed the acceptance of the scientific method, is called the Scientific Revolution. The Scientific Revolution, in its turn, laid the groundwork for the industrial revolution. These two revolutions, scientific and industrial, generated a cascade of changes to human existence, comparable to the rise of agriculture, 10 millennia earlier.

Consider what has been achieved through the progressively more clear understanding that the scientific method has given us. Gradually perfected scientific theories have allowed us to navigate the globe, fight diseases, harness electricity, improve crop yields, synthesize fabrics, fabricate computer chips, send out probes to explore our solar system, and share cute cat videos on the internet.

Scientific theories are our best description *at this time* of how the universe works.

When you think back on Aristotle's theory about how stuff falls down, the most important thing that he contributed may be the very idea that you *can* make a theory about how the world works. You don't have to just say "The gods made it that way" and walk away.

Aristotle's pronouncements showed that you can make a theory and then see how accurate and helpful that theory is. The Scientific Method gives a disciplined set of steps to go through, which weed out ideas and hypotheses that are contradicted by real-world experience, leaving only theories that are well supported. And to be considered truly worthy, a theory must predict unique phenomena that can be tested, and shown to be accurate (or not).

Due to the fact that we are dealing with a universe that exists on scales ranging from the mind-bogglingly vast to the mind-numbingly small, a disciplined method for guiding our understanding of reality has proved invaluable.

And speaking of mind-bogglingly vast and mind-numbingly small, science also developed a way of handling those crazy-big and crazy-small numbers. We look at **scientific notation** next.

Review

The essential foundations of modern science are rational thinking, critical thinking, and empiricism.

"Rational thinking" means that you check to see that your thoughts don't contradict each other. "Critical thinking" means that you verify your thoughts and beliefs against real-world data. "Empiricism" means that you have more confidence in real-world observations than in stories that you make up in your head (or that other folks put into your head).

To be fair, I should note that real-world observations can be a little tricky. A single observation, a single data point is not enough to go on. Anecdotal evidence (Joe said that Sally said that this is true) is not worth much. What you need are direct observations, and your set of observations had better be **statistically significant**. It gets technical, but follow this link if you want to see the details.[18] The Too Long Didn't Read version is that there are proven methods to check that your set of results is large enough, and consistent enough, to *almost* guarantee that you are not being fooled by random noise in the data.

There are also **logical fallacies** to watch out for. These are basically failures in rational or critical thinking. They can be very sneaky! There's a helpful list at this link.[19]

Building on the foundations of empiricism, rational thinking, and critical thinking the **scientific method** makes the outrageous demand: "Show me." Whatever your hypothesis is, it had better make some unique predictions that can be tested with current technology. Otherwise, it's just hot air.

The scientific method is not about The Truth. It's not about what you *want* to believe. It's about weeding out ideas that don't match reality, until you are left with theories that accurately predict what actually happens in the world around us.

Long story short: the scientific method says "Show me!"

Activity

Is That Idea Falsifiable? (Does It Make Predictions That You Can Test)

A hypothesis isn't worth anything if you can't test it.

This activity might be easier if done with a partner, a helper, or in a group.

Pick a hypothesis that you think is wrong. If nothing comes to mind, then take the hypothesis that "the Earth is flat." Or, if you believe that the Earth is flat, then take the hypothesis that "the Earth is a sphere." You are looking for a hypothesis that makes predictions that you believe can be disproven by real-world tests.

Write down one or more predictions that this hypothesis makes.

Now write down how to test each of these prediction(s) in the real world.

Do you think that those test(s) will support or disprove the hypothesis?

I take on the hypothesis that "Earth is flat" in this video:
https://www.whatacuriousworld.com/chapter-3/#Video03

Links

Biography

16. "Francis Bacon (Stanford Encyclopedia of Philosophy)." 29 Dec. 2003, https://plato.stanford.edu/entries/francis-bacon/.

Biography

17. "Freeman Dyson - Wikipedia." https://en.wikipedia.org/wiki/Freeman_Dyson.

Physics-Speak

18. "Statistical significance | Definition, History, & Facts - Britannica." 26 Aug. 2023, https://www.britannica.com/topic/statistical-significance.

Deeper Dive

19. "Logical Fallacies, University of Nevada." https://www.unr.edu/writing-speaking-center/writing-speaking-resources/logical-fallacies.

4. Scientific Notation

"Space … is big. Really big.
You just won't believe how vastly, hugely, mindbogglingly big it is.
I mean, you may think it's a long way down the road to the chemist's,
but that's just peanuts to space."
~ Douglas Adams

Don't panic. This is one of the shortest chapters in this little book!

While I have made a sincere effort to avoid unnecessary math stuff, mind-bogglingly big (and small) numbers have a way of creeping in when you are talking about this wacky universe that we inhabit. There is one tool that I think helps clear the mental fog when we talk about ridiculously huge or unimaginably tiny stuff.

I'm going to introduce an old friend of mine called **scientific notation.**[20] Scientific notation works by just giving you the first few "significant" digits of a number, and then telling you how many zeros you have to tack on in order to get the full effect.

For example, there are about 1,670,000,000,000,000,000,000 water molecules in an average drop of water. Who wants to count all of those zeros? This same exact number could be written as "1.67×10^{21} molecules" in scientific notation.

The "10^{21}" part means "multiply 1.67 by 10 and repeat multiplying that by 10 for a total of 21 times." Or you could think of it as "move the decimal point 21 places to the right." The little number next to "10" is called an exponent, and it can be positive (move the decimal point to the right) or, for tiny fractions, it can be negative (move the decimal point to the left).

In scientific notation you just use the first few digits of your measurement. In the above example that would be "167". You write the first digit "1", then a decimal point ".", then the rest of the digits to the right of the decimal point "67". In the above example that gives "1.67". Finally you add the " $\times 10^{\text{whatever}}$" where "^{whatever}" tells the reader how many digits (right or left) they need to move the decimal point. In the above example, you have to move the decimal point 21 places to the right to get back to the longhand version of the number, so the result in scientific notation is "1.67×10^{21} molecules".

Let's walk through some exponents of 10 for the number 1.67:
$1.67 \times 10^{1} = 16.7$
$1.67 \times 10^{2} = 167.0$
$1.67 \times 10^{3} = 1{,}670.0$
$1.67 \times 10^{4} = 16{,}700.0$
$1.67 \times 10^{5} = 167{,}000.0$

…
$1.67 \times 10^{10} = 16{,}700{,}000{,}000.0$
…
$1.67 \times 10^{20} = 167{,}000{,}000{,}000{,}000{,}000{,}000.0$
$1.67 \times 10^{21} = 1{,}670{,}000{,}000{,}000{,}000{,}000{,}000.0$ That's the number we are aiming for!

So the "1.67×10^{21} molecules" is just a very compact way of writing the number of water molecules in an average drop of water. You can tell by the "10^{21}" that this is a humongous number. The "10^{21}" gives you the magnitude of the number.

Scientific notation makes it easy to compare two numbers. For example, 2×10^7 is greater than 8×10^5 because *anything* with *7* zeros behind it is greater than *anything* with *5* zeros behind it. It's another way of saying that 20,000,000 is greater than 800,000.

When you are dealing with numbers with up to 5 or 10 zeroes in them, you probably don't need scientific notation. But when there are more zeros than that, scientific notation's little exponent next to the "10" means that you don't have to count up all of those zeros.

For fractional numbers such as 0.0054, you use a minus sign on the exponent to show that you are moving the decimal point the other way. In scientific notation that would be 5.4×10^{-3}. The -3 means "divide this number by 10 and continue to divide by 10 for a total of 3 times." Or to put it another way, "move the decimal point 3 places to the left." So 5.4×10^{-3} is the same as writing 0.0054 — and that becomes really handy when the numbers are mind-numbingly tiny. For instance, Planck length (which we will meet again in the appendixes) is 1.6×10^{-35} meters, or you could write that out longhand as 0.000000000000000000000000000000000016 meters. You *really* don't want to try to count out all of those zeroes.

I will be using scientific notation for mind-bogglingly big and mind-numbingly small numbers. I think it makes those numbers just a little easier to comprehend, and a heck of a lot easier to compare one humongous number to another humongous number.

When you see something in scientific notation, that little exponent number above the "10" is the most important part.

Now that that's behind us, in the next chapter we'll get back to the main event.

Activity

How Far Is It From Here to There (in *Scientific Notation*)?

What you'll need for this activity:

- A measuring tape, yardstick, or some such distance-measuring device or app.
- Something to scribble your work on (pencil & paper, a notes app, etc.)

Pick a distance between two parts of your home. Measure this distance in inches or centimeters. For example, in my house, it is 263 inches from the front door to the bathroom door. There have been times when this distance was very important to me.

To put that into scientific notation, you would move the decimal point over two places and write it as 2.63×10^2 inches. It's not like you need scientific notation with a small number like this, it's just for practice.

How many inches is the distance that you chose? How does that look in scientific notation?

Truth be told, scientific notation is not that useful for reasonably sized numbers. It even gets a little goofy for numbers smaller than 100. For instance, 99 becomes 9.9×10^1 (meaning: take 9.9 and multiply it by 10). Numbers smaller than 10 are even sillier! The number 5 gets written as 5.0×10^0 (meaning: take 5 and don't multiply it by anything, it's just fine the way it is). As you can see, using scientific notation for reasonably sized numbers is not only silly, it's a waste of paper. But seriously, folks. Scientific notation is really useful for ridiculously large numbers and ridiculously small fractions.

Link

Deeper Dive

20. "Scientific Notation - Math is Fun." https://www.mathsisfun.com/numbers/scientific-notation.html.

5. Weighty Subjects

"Gravity is a habit that is hard to shake off." ~ Terry Pratchett

So things fall down because they're heavy, right? Well, sort of.

Gravis is the Latin word for heavy. Our modern word gravity comes from *gravis*. We have several words in English that we use when we talk about stuff that falls down. The words weighty, heavy, and gravity come to mind. But these same words can also be used to describe something abstract like a serious topic or a situation with serious consequences. If you respect someone's opinions, you could say that their opinions carry some weight. A fancy way to say that a person is respected or influential is to say that they have gravitas. In my youth, I might have said that an intriguing statement (e.g., *"What makes night within us may leave stars." ~ Victor Hugo*) was really heavy. The experience of weight/gravity/heaviness is so powerful in our physical experience that we use these same words to describe ideas and events that have an impact on us or on the world — even when that impact has nothing to do with physical weight or gravity, nothing to do with things that fall-down-go-boom.

So you can see that in everyday speech these words can either mean a physical experience, or they can mean that some person/thought/event has an important but non-physical impact on the world.

A word that seems straightforward will have different meanings depending on the context in which it is used.

Massively Important

In order to nail down how the universe works, we need to use words in well-defined ways — regardless of how those words get bent to other uses in daily speech. "Mass" is one such word. When you think of something massive I bet you think of something large and heavy. Something very hard to move. That "very hard to move" part is what physicists are talking about when they use the word mass,[21] and to understand the first really successful theory of gravity you have to wrap your mind around the physicist's idea of mass.

It Takes Force To Mess With Inertia

Another word that gets used for non-physical activities as well as physical objects is "inertia." You might say that a bowling ball that is rolling down a hill has a lot of inertia, meaning that it would take a lot of force to stop it in its tracks. You could also say that some political candidate who's polling numbers just keep getting better and better also has "inertia." And here you also mean that it would take a lot to stop them in their tracks.

When I say inertia in this little book, I'll be talking about physical objects — not political campaigns. The physical meaning of inertia is that it takes force to change what an object is doing.

If a bowling ball is sitting still, it will take some force to get it moving. The inertia of a bowling ball that is sitting still keeps that bowling ball sitting still. If you want to get it moving, it will take some force to overcome the bowling ball's "sitting still" inertia.

If a bowling ball is rolling down a bowling lane, it will take some force to stop it. The inertia of a rolling bowling ball keeps that bowling ball rolling along. If you want to stop it in its tracks, it will take some force to overcome the bowling ball's "rolling down the lane" inertia.

Physical objects *always* have inertia — whether the object is sitting still or moving. Inertia keeps an object doing exactly whatever it's been doing. If an object is sitting still, inertia keeps it sitting still. If an object is moving, inertia keeps it moving in the same direction at the same speed.

If an object that is sitting still suddenly starts moving, something messed with it to overcome its "sitting still" inertia. Some force got it moving.

If a moving object suddenly changes speed or veers to one side, something messed with it to overcome its "going in a straight line at a constant speed" inertia. Some force pushed on it from one direction or another.

If this sounds a lot like my description of the physicist's definition of **mass** as being something that is "very hard to move," it should. To a physicist, the "sitting still" **inertia** of an object *is* its mass. In fact, a physicist calls this "sitting still" inertia an object's **rest mass**. If you eliminate friction, and you measure how much force it takes to get a "sitting still" object moving, you have measured the amount of inertia it has. This amount of inertia *is* the object's mass.

So far, we've talked about mass and we've talked about inertia. Now let's talk about weight.

Mass Is Not Weight

As of 2024, the International Space Station (ISS) has been in orbit and continuously occupied for 24 years. It is massive, and yet while in orbit it has no weight.

If it were resting on a sandy beach in Florida, it would weigh over 900,000 pounds. But it isn't resting at sea level, it's about 250 miles above those sandy beaches and travels at right around 17,500 mph,[22] circling the Earth. It and the people in it are weightless. How can the ISS have such a large mass and be weightless at the same time?

The ISS is in orbit, continuously being pulled down by the Earth's gravity. It is falling, and yet it never falls to Earth. To understand how that is possible you need to mull over some thought experiments about how things move and fall.

A **thought experiment** is a way to take things that you know, and then to imagine what would happen in different circumstances. A thought experiment can be useful in understanding how something out of the ordinary could happen. For instance, how the ISS could be in orbit. And what "being in orbit" actually means.

There are a couple of cautions to be aware of with thought experiments. First, if your thought experiment does not suggest ways to test its predictions in the real world, it isn't worth much. Second, it's easier to fool yourself when you are making it all up in your head — so be extra vigilant for things that violate rational or critical thinking.

Let's start by imagining that you are standing in a huge open field and are throwing baseballs horizontally. If you toss a ball gently, it doesn't go very far before it hits the ground. If you really lean in and throw hard, the ball will travel much farther before it hits the ground. The faster the ball goes, the farther it gets before Earth's gravity brings it down to the ground.

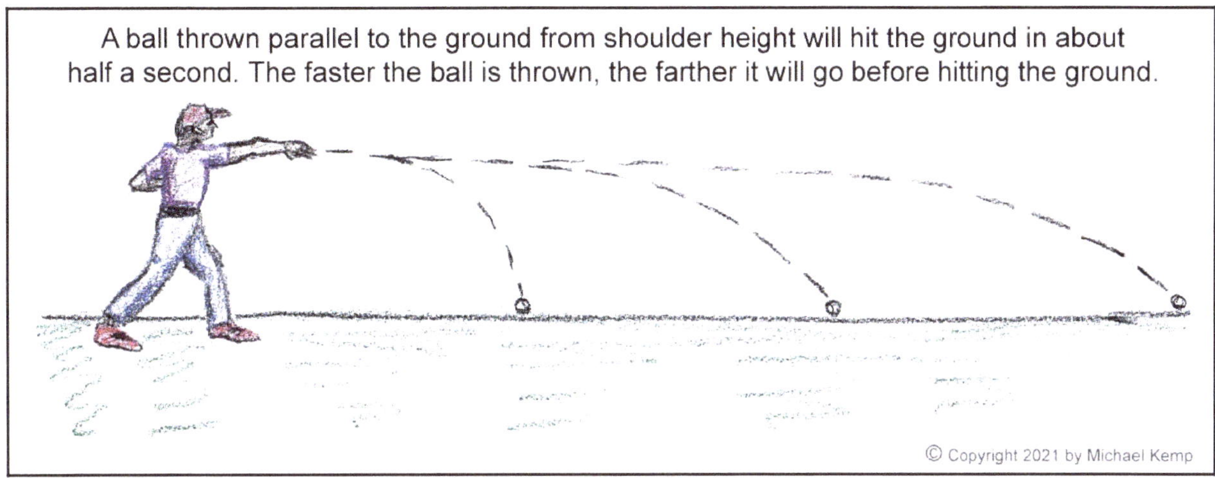

A ball thrown parallel to the ground from shoulder height will hit the ground in about half a second. The faster the ball is thrown, the farther it will go before hitting the ground.

To an observer standing to the side with a stopwatch, each horizontally thrown baseball will hit the ground in the same amount of time regardless of whether it was thrown slowly or at great speed. The gravity of the Earth brings each baseball to the ground at the same rate, but the faster the baseball is thrown, the farther it travels in that fixed amount of time.

An Orbit Is Gravity Balanced by Inertia

Now imagine yourself up in space above the Earth's atmosphere, in a space suit with a rocket pack that keeps you from falling down to Earth. From there you can see the curvature of the Earth. If you were to simply let go of a baseball, it would be pulled straight down by Earth's gravity.

If from your spot in space, you toss a baseball slowly in a horizontal direction it will quickly curve down. It won't go far forward before heading on down to the Earth. The combination of the forward movement of the ball and the pull of gravity will make the ball follow a curved trajectory.

With a slow pitch, the ball doesn't travel very far ahead of you as the seconds tick by.

With a fast pitch, the ball travels much farther ahead of you as the seconds tick by.

Regardless of the speed at which the baseball is traveling, gravity pulls on it with the same force, but because the fastball travels farther forward as the seconds tick by, the fastball will have a flatter trajectory than the slow ball. The faster the ball is thrown, the flatter the curve of its flight.

Now imagine that you are 250 miles above the Earth and that you can propel the baseball at 17,500 mph. That baseball would still fall toward the Earth, but it would be moving so fast horizontally that the curve of its fall would be almost flat.

At 250 miles altitude and 17,500 mph, the curve of the baseball's trajectory would just match the curvature of the Earth.

It would not fly straight out into space because the Earth's gravity is pulling on it; it would not hit the ground because the Earth is a sphere and as the baseball falls, it will have traveled just far enough forward to stay at the same height above the curvature of the Earth.

Viewed from the side, the baseball's falling path would trace a circle all the way around the Earth, as is shown in the illustration below.

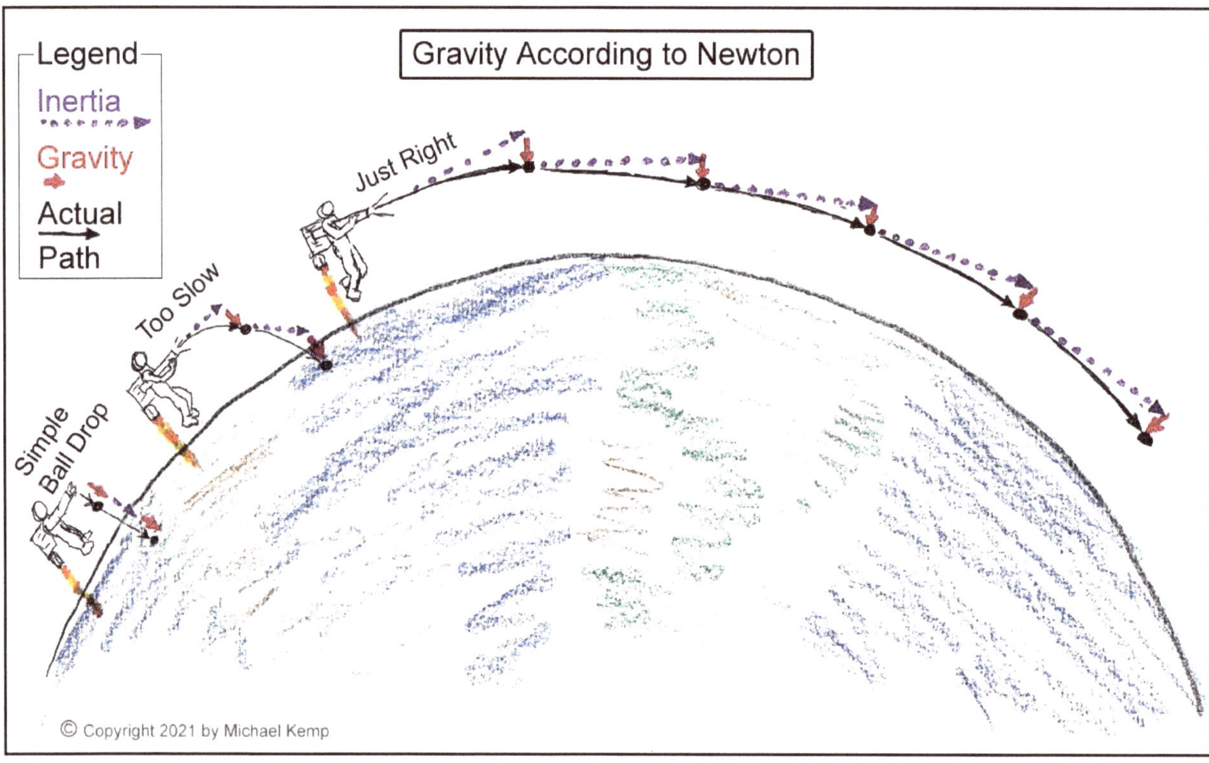

Note that this has nothing to do with the rotation of the Earth. In this illustration, the Earth could be rotating clockwise, counter-clockwise, or top-to-bottom. The Earth's gravitational pull on the ball would not change in any meaningful way regardless of the direction in which the Earth is rotating.

This is what it means to be in a stable circular orbit. Any slower at that altitude and the ball would curve down faster than the curvature of the Earth and it would fall to the ground (if it didn't burn up in the atmosphere). A faster speed at that altitude and the Earth's gravity would not pull it down fast enough to keep it in a circular orbit and it would loop farther out. For any object at the same altitude as the ISS, a speed of 17,500 mph is perfect to keep that object in a stable circular orbit, falling around the Earth every hour and a half, just like the ISS.

As an aside, the 250-mile altitude and 17,500 mph orbit is just one example of a stable circular orbit around the Earth. The pull of Earth's gravity gets weaker as you go farther out, so satellites that are in a higher orbit travel at a slower speed to achieve a stable circular orbit.

The ISS is in about the lowest practical orbit due to the problem of atmospheric drag as you get closer to the surface of the Earth.[23] Any lower and it would be gradually slowed down by the friction of moving through the top layers of Earth's atmosphere. Not good.

Below are some examples of things in orbit around the Earth. Note that the farther out you get, the less speed you need to maintain a stable orbit.

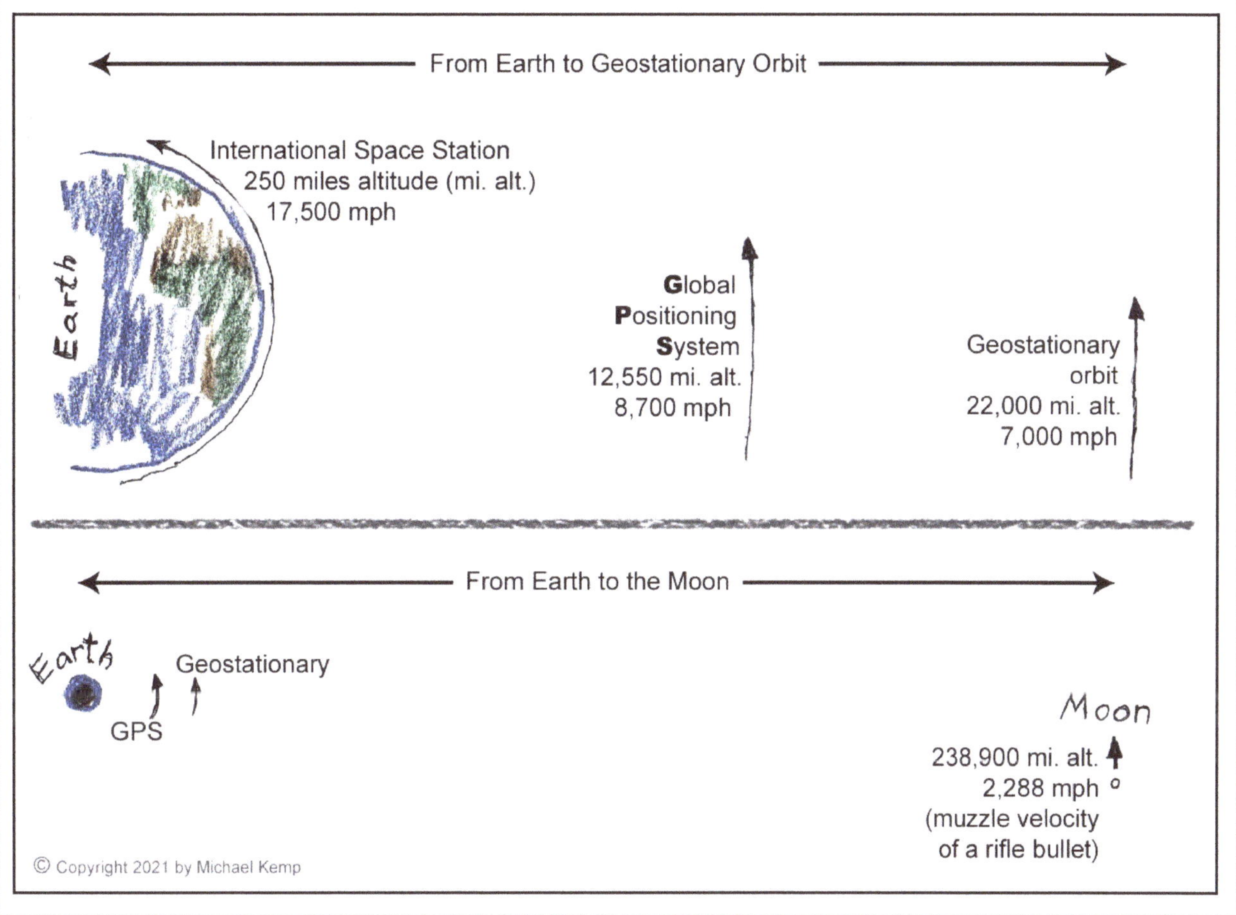

So I repeat the question: how could the ISS have a weight of 900,000 pounds if it were sitting at sea level and be weightless in orbit? Well here, my friend, is the difference between mass and weight. It actually still has a mass of 900,000 pounds; it's just perpetually falling around and around our planet, traveling forward just fast enough that it never actually hits the ground. A falling object does not experience weight. If you had some monster-sized weight scales to slip under the ISS, the scales would be falling right along with the ISS. The scales would read zero pounds. Weightless.

Another way to think about mass versus weight is to imagine the ISS in various places. It would weigh 900,000 pounds if it were lying on a sandy beach on planet Earth. But now try a thought experiment where the ISS is way out in deep space, far away from any star, with no detectable gravity. It has no "down." It has no weight. Let's imagine that as far as we can tell, it is sitting very still out there in deep space. Could you move it by pushing on it? That would take a lot of force. Why? Because even though it is weightless it still has 900,000 Earth-pounds of mass. 900,000 Earth-pounds of inertia.

If you want to see another description of inertia and mass, and some notes about how Galileo was instrumental in developing those concepts, follow this link.[24]

How much something weighs and how much mass it has are directly related, but not necessarily in the way that common sense would have us believe. Our experiences in life give us our common sense. I think it's safe to say that all of us have, at some point, dropped a weighty object on our foot and have regretted it. Common sense tells us that weight is a big deal, so we tend to think that the weight of an object tells us whether it is massive or not. If you want to understand the mechanics of the universe, you sometimes have to let go of common sense. It's not an object's weight that determines its mass, it's an object's mass, plus that object's location, which determines its weight. An object's weight varies depending on the gravitational field that you put the object in. Weight, like the direction of "down," depends upon location. The object's mass would be the same whether it was on Earth, Mercury, Jupiter, or floating out between the stars. But that same object's weight would vary depending on its location. An object that weighs 100 pounds on Earth would weigh only 38 pounds on Mercury (lower gravity), 240 pounds on Jupiter (higher gravity), and essentially nothing out between the stars (microgravity). Mass combined with gravity creates weight. Mass is *the* big deal, weight varies. The common sense shortcuts that serve us so well in daily life often obscure what is really going on in the universe.

Truth be told, mass is kind of a crazy thing. Why does stuff even have mass and inertia in the first place? If you are curious, here are a couple of video links to get you started. Fair warning: these are deep waters.[25] [26]

You have to wrap your mind around the physicist's idea of **mass** if you want to understand the first truly successful theory of gravity. I hope your mind is primed because a guy named Issac Newton completely changed the game by changing the question.

Instead of asking "*How* does stuff fall down?" as Aristotle and Galileo did, Newton asked himself "*Why* does stuff fall down?" That is a much more powerful question, which yielded a much more powerful answer.

Review

In this chapter we saw how *weight* is dependent on an object's *mass* and the *gravitational field* that the object finds itself in. An object's mass is the same regardless of where it is. An object's weight varies depending on its location. If it's someplace with a big gravitational field (Jupiter, for instance) it will have a lot of weight. If the same object is someplace with a small gravitational field (Mercury, for instance) it will have a much smaller weight. If it's in deep space between galaxies (microgravity) it will have no weight. And in each location, it will still have exactly the same mass.

An object's mass plus the gravitational field that it is in *creates* its weight. As with the direction of down, an object's weight depends on its location. Its mass stays the same, its weight varies by location.

The inertia of an object — how much force is needed to change its speed or direction — is the same as that object's mass. Even an object sitting still has inertia. It takes force to mess with it, to get it moving.

Satellites in orbit around the Earth make use of their inertia to keep them from falling to the ground. That's true for everything from manmade objects like the International Space Station and GPS satellites to natural objects such as the Earth's Moon.

At the beginning of this chapter I noted that we re-use words about physical weight when we talk about influential people or ideas. In the same way, when I have an attitude or opinion that I have put a good deal of energy into, I've noticed that it has something like "inertia." That attitude or opinion really resists being changed by new information that pushes on it. This is just an analogy, saying that a closely held attitude or opinion has its own kind of inertia, but it's worth thinking about.

Activity

Let's Play With an Orbit

It's easy to simulate how the force of gravity versus the inertia of a "satellite" keeps the satellite in a stable circular orbit.

What you'll need for this activity:

- Two or three sheets of paper (regular 8-½" x 11" printer paper works well).
- 6 or 8 feet of string or twine.
- Some tape.

Crumple the papers up together into a single tight ball. Tape one end of the string to the ball, leaving something like 6 feet of string hanging off your paper ball.

Go somewhere where you've got plenty of room to swing this paper ball around by the string without knocking down any lamps or houseplants or whatever. A lawn or a park would be a good choice.

Swing the paper ball around your head in a big circle.

If you let go of the string, observe how the paper ball goes winging off in one direction. This is what the inertia of the paper ball always *wants* to do.

Grab the string and start swinging the paper ball around your head again. Feel the tension on the string that keeps the paper ball from winging off in a straight line. That tension on the string is just like the pull of gravity that keeps satellites (like the ISS or the Moon) from winging off into deep space.

You have just simulated a satellite in orbit around the Earth. You have also demonstrated inertia (when you let go and the paper ball went sailing off in a straight line). And the tension on the string while you are swinging the paper ball around your head? That is like the force of gravity, keeping the satellite in orbit.

And here I am on our back porch playing with string and paper:
https://www.whatacuriousworld.com/chapter-5/#Video05

Links

Physics-Speak
21. "Mass | physics | Britannica."
https://www.britannica.com/science/mass-physics.

Deeper Dive
22. "What Is the International Space Station? - NASA." 30 Oct. 2020,
https://www.nasa.gov/audience/forstudents/5-8/features/nasa-knows/what-is-the-iss-58.html.

Deeper Dive
23. "Comparison satellite navigation orbits.svg - Wikimedia ...."
https://commons.wikimedia.org/wiki/File:Comparison_satellite_navigation_orbits.svg.
Be sure to click the illustration to get the animation going.

Deeper Dive
24. "Inertia and Mass - The Physics Classroom." 31 Jan. 2001,
https://www.physicsclassroom.com/class/newtlaws/Lesson-1/Inertia-and-Mass.

Great Video
25. "It's Not What you Think. Where does mass come from? YouTube." 21 Sep. 2018,
https://www.youtube.com/watch?v=2kUFs6_DBrM.

Great Video
26. "PBS Space Time - The True Nature of Matter and Mass YouTube." 6 Jan. 2016,
https://www.youtube.com/watch?v=gSKzgpt4HBU.

6. Newton Asks "Why?"

"I think, at a child's birth, if a mother could ask a fairy godmother to endow it with the most useful gift, that gift should be curiosity." ~ Eleanor Roosevelt

It's a good thing, a useful thing, to have reliable theories about the way the universe works. You can think of a scientific theory as an educated guess at what the real rules of the universe are. The more accurate a theory's description of the way the universe works, the more dependable its predictions will be. Test those predictions and you test how closely the theory matches the way our universe is actually built.

We've seen that Galileo made a pretty good theory about how stuff falls down.

"How" only tells you the observable part of an event. "Why" tells you about the underlying rules of the universe that caused that event to occur. It's a whole level up to have an accurate theory about *why* stuff happens, rather than just a theory about *how* stuff happens. Actually, it's not just a level up, it's a whole different game. A much more powerful game.

Isaac Newton took it from "how" to "why."

In general, Newton led a privileged life, but he had a rough start.[27] His recently widowed mother bore him prematurely in 1643. She remarried when he was three years old, and they sent him to live with his grandmother. He loathed his stepfather and was alienated from his mother. Eventually, when his stepfather died, Isaac moved to the estate and tried to manage the family holdings. It became obvious to everyone that he was not suited to the task. Newton, thankfully, returned to grammar school and then went on to pursue higher education.

At Cambridge, he earned his way by performing valet duties until he was awarded a scholarship. The Cambridge philosophical education at that time was based on Aristotle's teachings, but on the side, Newton dug into the somewhat heretical ideas of Kepler, Galileo (who had died just the year before Newton was born), Bacon, and other upstarts. After some time at Cambridge, the school closed to avoid spreading the contagion of what would turn out to be Britain's final round of the bubonic plague.

During the hiatus, while the school was closed, Newton retreated to the family farm. There he went off the scholastic deep end, embracing the Scientific Revolution[28] and developing his own ideas in mathematics, optics, laws of motion, and (yes) gravity — based on the ideas and attitudes of Copernicus, Descartes, Galileo, et al. Not Aristotle.

When the plague was over and Cambridge reopened, Newton returned to the school and was soon given positions of respect and responsibility — although there was some concern about his unorthodox religious and philosophical ideas. This escalated to the point that Charles II, King of England, Scotland, and Ireland had to step in and grant him a special exemption from conformity to some of the Cambridge traditions.

Newton's groundbreaking ideas, developed using the principles of the Scientific Revolution, proved to be so trustworthy that they soon became accepted as full-blown *Laws of Nature*. You could calculate ballistics with unprecedented accuracy, account for ocean tides, and under one set of consistent theories, understand everything from colliding billiard balls to the motions of the planets.

And that thought experiment that I used in the last chapter about throwing a baseball into orbit around the Earth? I got that from Newton. He used that thought experiment, in a book published in 1726. The thought experiment is called "Newton's Cannonball"[29] in case you want to look it up. His version involves firing a cannon from the top of an imaginary mountain whose summit is above the Earth's atmosphere. I just changed it to be an astronaut firing off a baseball to be a little more in tune with the times.

Isaac Newton's theories have had a huge and lasting impact on modern civilization. As a matter of fact, they paved the way for me to be composing this very document. I started writing this book using an internet app. At that time, my connection to the internet was through a satellite dish out in the meadow next to our house. It pointed to a satellite that orbited the Earth at a distance (approximately 22,000 miles) such that the time it took for that satellite to complete one orbit of the Earth was exactly one Earth day. It orbited in the same direction that the Earth rotates so that it appeared to those of us on the ground as if it were at a fixed point, hanging in the sky. This is called a **geostationary orbit** because a satellite parked in that orbit stays in one spot (stationary) in our sky.

That satellite could not have been conceived of without Newton's Cannonball. Thanks, Isaac! The idea of a geostationary communication satellite was popularized much later (in 1945) by a science fiction writer named Arthur C. Clarke.[30] Thanks, Arthur!

So, Newton created a theory of gravity so good that it's used for calculating everything from artillery ballistics tables to geostationary orbits, to the orbits of Jupiter's moons. How in the world did he manage to do that?

Remember that bit about Cambridge closing for the plague and Newton going back to the family farm for a couple of years? If you have ever heard of Isaac Newton being conked on the head by a falling apple and having a "Eureka! Gravity!" moment, it would have been during those two years while Cambridge was closed. Except for the little detail that he didn't say that he had gotten conked on the head. What he *did* say was that one warm day after dinner he went into the garden and drank tea under the apple trees. He was in a contemplative mood when an apple happened to fall to the ground. That started him pondering why apples (and other objects) fall straight down. From that, he gradually developed his own theory of gravity. Not about how stuff falls down but *why*. Why would an apple fall to the ground?

Newton's observation of the falling apple was more of a "that's funny" moment and not so much a "eureka!" moment. Newton acknowledged his reliance on the works of those who went before him by declaring *"If I have seen further it is by standing on the shoulders of Giants."* Galileo was one of those "giants," and one of the critical ideas that Galileo developed in regard to how objects move was the concept of inertia. You may recall that I mentioned how **inertia** and the physicist's concept of mass are directly connected. In fact, they will use the term *inertial mass* to mean the basic, resting mass of an object. But I didn't tell you much about inertia other than that inertia is how much force is needed to get an object moving, or to stop a moving object in its tracks.

The *force* needed to overcome the inertia of an object is calculated by multiplying an object's inertial mass by the amount that you accelerate that object. The formula for this is **F=ma** where "**F**" is the force needed to "**a**" accelerate an object with a mass "**m**". That's all nerd-speak, so let me give you a couple of ballpark examples of inertial mass and the force needed to accelerate that mass.

Let's say you and a friend are out in a field tossing a baseball back and forth. When you are holding the baseball in your hand, it won't go anywhere until you put some effort into throwing it. That is to say, until you apply an external and unbalanced force, to use Newton's way of speaking.

When you throw a *slow* ball to your friend you are making a small acceleration to the inertial mass of the baseball. Your hand can feel that you have to use a little bit of force to throw the baseball. When your friend catches the ball, his hand will feel a small amount of force from the inertial mass of the baseball as he decelerates it to a stop. So, a small acceleration or deceleration of a fairly small inertial mass (the baseball), takes a small amount of force.

When you throw a *fastball* to your friend you are making a large acceleration to the inertial mass of the baseball. Your hand can feel that it takes more force to get the baseball going really fast. When your friend catches *that* ball, his hand will feel a large amount of force from the baseball's inertial mass as he decelerates it to a stop. Ouch. A large acceleration or deceleration requires a large amount of "external and unbalanced" force.

What happens if you throw a slow pitch with a bowling ball? Now we're talking about a much larger inertial mass. If you throw a slow pitch with a bowling ball, you can bet you'd *really* feel that throwing the bowling ball takes a *lot* of force. When your friend catches the bowling ball, his hand will *definitely* feel it! A big mass requires a big force to accelerate or decelerate it.

Newton's **F=ma** formula says that, with either a big acceleration **a** or a big mass **m**, you need a big force **F** to mess with an object's inertia.

Galileo often combined his study of inertia with his study of gravity. Inertia works to keep objects in their current state. Inertia keeps stationary objects stationary, and it keeps moving objects moving. Gravity, on the other hand, works to get objects moving toward the center of gravity.

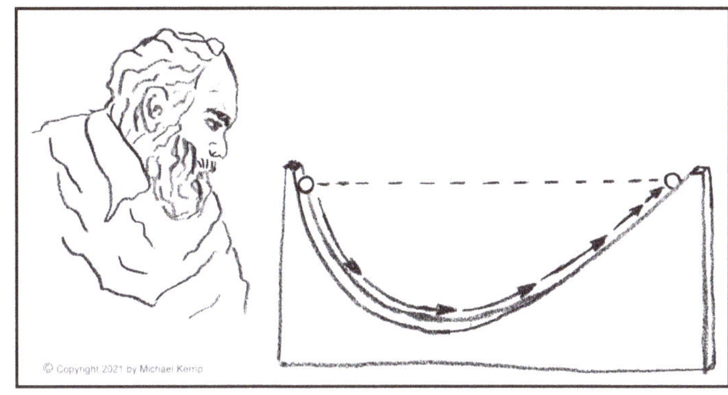

Galileo did a lot of experiments with balls rolling down ramps at various angles.[31] He used a water clock, strategically placed bells, and careful observation of the distance a ball rolls down an inclined plane in given time intervals to calculate the acceleration of gravity. He also observed that the inertia of a ball released down one incline would cause it to return to the same height up another incline, regardless of whether the angle going down matched the angle going up or not.

So that's a bit more background on what it means when we speak of inertia. Getting back to Newton's apple, we have an apple whose stem has just separated from the branch. Inertia works to keep the stationary apple stationary, but gravity works to accelerate the apple to the ground. What exactly is this force of gravity that overcomes the apple's inertia? A big clue is that this force accelerates the apple straight down to the Earth.

Aristotle would have said that the apple, being composed of the earth element, was simply seeking its proper sphere: earth.

Newton wasn't buying it.

Galileo described *how* the apple fell but with no explanation of *why*.

At the same time that Newton was puzzling over gravity, he was also working out what are now called Newton's Laws of Motion,[32] and inertia plays a prominent role in those Laws.

Newton set down his three Laws of Motion, and his law of gravity, in a book that is generally referred to as *Principea*, published in 1687.

Newton's Laws of Motion

Newton's First Law of Motion is **An object at rest will remain at rest unless acted upon by an external and unbalanced force. An object in motion will remain in motion unless acted upon by an external and unbalanced force.**

Here's another way to put it: A baseball is not going to leap out of your hand for no reason. It will sit there until you push on it. Once you throw a baseball it will continue moving in the direction you threw it (its course adjusted by the gentle forces of air friction and gravity) unless somebody catches it or hits it. If the batter clobbers it, the baseball's direction and speed will change. It has been accelerated! The bat has applied an external and unbalanced force to the ball, to say the least.

As a thought experiment, I hope you will agree that if a baseball were thrown in the frictionless vacuum of space, far away from any source of gravity, once it leaves your hand it would continue at a constant speed in a straight line. And it would continue at that constant speed in a straight line until something interfered with it. Anything that interfered with it would be an "external and unbalanced force" in Newton's way of speaking.

Returning again to Newton's apple, the first Law of Motion says that just because the stem separates from the branch is no reason for the apple to fall — *unless* some external force is acting upon it. Inertia would keep the stationary apple stationary, suspended in mid-air… unless some external and unbalanced force acted upon it.

Newton's Second Law of Motion states that for a specific amount of force, the more massive an object is the less it will be accelerated. Push a little on a baseball, it moves. Push the same amount on a big boulder, not so much. So if there were an equal force applied to objects of different masses, the object with the smaller mass would be accelerated more than the object with the greater mass. The folks at NASA put Newton's Second Law of Motion this way: **The acceleration of an object depends on the mass of the object and the amount of force applied.** This is where that **F=ma** formula comes from. You can flip the equation around to solve for acceleration and get **a=F/m** which says "acceleration equals force divided by mass." The more force you use, the greater the acceleration. The more mass an object has, the less acceleration you are going to get.

If you take Newton's first two Laws of Motion together, they add up to what we call inertia. The first Law says that it takes some amount of force to get an object to accelerate (or decelerate). The second Law acknowledges that you need more force to accelerate more massive objects.

The more massive an object is, the greater its inertia. Less massive objects have less inertia. The planet Earth has a lot more inertia than an apple. An apple is easier to interfere with than the Earth. Neither you nor I can disturb the inertia of the Earth to any detectable degree. On the other hand, most folks could pick up an apple and throw it. The more massive the apple, the more effort would be required to get the apple moving. But the Earth is just plain out of our league.

Newton's third Law of Motion is that **whenever one object exerts a force on another object, the second object exerts an equal and opposite force on the first.** This is sometimes stated as **every action causes an equal and opposite reaction**.

An Apple, Inertia, and an External Force

When I muse about how Newton developed his theory of gravity, it seems to me that it parallels his thinking about his Laws of Motion. In my mind's eye, I can see Newton wondering: "To get that apple to fall down, what was acting upon that apple to make it fall? Why did the apple move when it separated from the tree branch? Why didn't the inertia of the apple just keep it hanging in mid-air?" This seems to me to be the way that Isaac Newton went from *how* stuff falls down to *why* stuff falls down.

His answer was: because some force was being applied to the apple. The really stunning aspect of his thinking is that he took apples, planets, the Moon, the Earth, and the Sun all into his mental cauldron and let them simmer on low heat. This was *not* a "eureka moment." It was the fruit of years of contemplation. If he were to make a theory about gravity to explain why the apple fell, it must also explain why the Moon is attracted to the Earth and orbits the Earth. And why the Earth and other planets are attracted to the Sun and orbit the Sun. He considered the apple, the Earth, and the Sun as the same except for the magnitude of their masses. His answer for a force that binds them all to each other was a new theory of gravity.

Newton expressed his theory of gravity as **Every mass attracts every other mass with a force that is directly proportional to multiplying their masses together, and inversely proportional to the square of the distance between them.**[33]

According to Newton, any objects that have mass attract each other. The bigger the mass, the more the attraction. And it works both ways. Massive object A attracts massive object B, and massive object B also attracts massive object A. How much do they attract each other? Do you add A's mass plus B's mass together to find out how much they attract each other? No. You *multiply* the mass of A times the mass of B to find out how big their mutual attraction is.

Also, the attraction of massive objects extends all the way across the entire universe. But, the further away two massive objects are from each other, the smaller the attraction. How fast does this gravitational attraction between massive objects decrease with distance? Pretty fast. "Inversely proportional to the square of the distance between them" sounds like math gobbledygook, but it's the same "inverse square law" that we experience in other parts of our daily life. Take how loud a sound is. If I'm standing right next to a person who is speaking in a quiet tone of voice, I can hear them pretty well. If I back off 10 feet, I'll probably still hear them OK, but their voice will sound much quieter. If I back off to 20 feet, I'll probably start having trouble hearing parts of sentences. At 30 feet, I'll catch a few words. At 40 feet, I won't be able to hear what they're saying at all. The power of their voice, which is strong at close range, drops off quickly as I get farther away. "Inversely proportional to the square of the distance between them" in Newton's law of gravity is just nerd-speak to say that the attraction between two massive objects drops off quickly with distance.

Even a huge attraction between massive objects (when they are close together) becomes very weak at astronomical distances.

And, of course, there are not just two objects in the universe that have mass. According to Newton's theory, every object in the universe that possesses mass is continuously attracting every other object that possesses mass. Even a speck of dust has mass and exerts a gravitational attraction on other objects that have mass — it's just that a speck of dust has so little mass that its gravitational force is smaller than the force of air friction. The gravity of a speck of dust is so trivial as to be meaningless in daily life.

> If something has no mass at all, such as a beam of light, it is not affected by gravity, according to Newton's theory. Spoiler alert: this little tidbit plays a critical role when comparing Newton's theory of gravity to Einstein's theory of gravity.

Looking back at apples falling in the garden while Newton sips his tea: Newton says that the force applied to an apple, that makes it fall, is gravity. Newton's theory of gravity states that an apple and the Earth attract each other with a force proportional to the mass of the Earth multiplied by the mass of the apple. They exert this force equally upon each other, so why does the apple fall to Earth while the Earth remains still? The force of gravity is an attraction. The Earth and the apple are applying a force to each other, attracting each other. When two objects attract each other, the *force* of attraction is "equal and opposite," as stated in Newton's Third Law of Motion. The direction of the Earth's attraction to the apple points "up," while the direction of the apple's attraction to the Earth points "down," so the directions are opposite.

The big difference is that (per Newton's Second Law) the Earth and the apple are accelerated by this *equal force of attraction* based on their respective masses.

And here we come to the issue of the relative masses of the apple and the Earth.

- An average apple has about a third of a pound of mass, which can be written as the decimal number 0.33 pounds. Or in scientific notation:
  3.3×10^{-1} pounds
- The Earth has about 13,200,000,000,000,000,000,000,000 pounds of mass.
  1.32×10^{25} pounds

Four apples would weigh around 1.32 pounds (1.32×10^{0} pounds). Compare that with 1.32×10^{25} pounds for the Earth and you can see that the Earth's mass is about 10^{25} times that of four apples. It would take about 40 trillions of a trillion apples to be as massive as the Earth. No, you are not seeing double. As unimaginable as just one trillion apples are, it would take 40 trillion of those one trillion apples to match the mass of the Earth.

Newton's Second Law of Motion can be applied here to say that, while two massive objects will both be moved by the force of their gravitational attraction, each object (the apple and the Earth in this case) will be accelerated by that force in inverse proportion to their own mass. "In inverse proportion" here means that the greater the mass, the less the acceleration — the smaller the mass, the greater the acceleration. The Earth has 40 trillion trillion times the mass of the apple.

When the apple separates from the tree branch, the mutual attraction between the apple and the Earth can now accelerate each of them toward the other. The *acceleration* caused by the separation of the apple from the tree branch will be 40 trillion trillion times less for the Earth than for the apple — so don't expect to see the Earth move. On the other hand, the constant force of gravitational attraction between the two of them will accelerate the apple at increasing speed — adding an additional 32 feet-per-second to its speed every second until the apple hits the ground (or hits some other object, such as Isaac Newton's head, if you prefer that image).

This was the first coherent and highly accurate theory about *why* stuff falls down.

The strength of gravity at the surface of the Earth has been measured to be an acceleration of 32 feet-per-second every second. This means that if you drop an object from a great height, and somehow remove air friction, when you first open your hand and let it go, it starts at 0 feet-per-second. One second later it will have accelerated to 32 feet-per-second. At the end of the second second it will have accelerated to 64 feet-per-second. At the end of the third second it will have accelerated to 96 feet-per-second. And so on and so forth, its speed accelerating an additional 32 feet-per-second every second until it hits the ground.

Experimental results confirm this all over the Earth. Well, OK. There are slight variations. Depending on local landforms, and differences in the density of the interior and surface formations of the Earth, the measure of the acceleration of gravity varies from about 32.09 feet-per-second per second, to about 32.25 feet-per-second per second. We use "32 feet-per-second per second" as a good approximation.

Newton's theories were supported by experiments and were taken to be *The Laws of Nature*. They proved invaluable to understanding the movements of the comets, moons, and planets. They were utilized in mechanics, engineering, and warfare. Newtonian physics was not seriously challenged for two centuries.

The scientific method, established by Sir Francis Bacon in the 1500s, enabled Newton to build a rational, coherent, experimentally verified set of theories about how the world works. Never before had such a complete, dependable, and consistent enumeration of the "Laws of Nature" been set down. And they worked anywhere you chose to test them. Almost flawlessly.

However, the very discipline of the scientific method that had enabled Newton to craft his breathtakingly logical, lucid, and eminently practical description of nature, also contained the seeds of his worldview's downfall. The scientific method demands that if empirical evidence contradicts the predictions of a theory, that theory is disproved. It may still be useful in many instances, but it is not quite right in the head. If the theory depends upon or predicts things that experiments show to be wrong, the theory is flawed.

And boy howdy, the experiments performed in the 1800s did just that. Hold onto your hat, friends, because this gets really weird.

Review

The main things that I hope you take away from this chapter are:

- There's a big difference between having a solid theory about *how* something happens and having a solid theory about *why* something happens.
- Newton built on top of Galileo's studies of inertia and motion to create his own "laws" of motion:
  - First Law: An object at rest is going to stay that way unless something messes with it. An object in motion is going to keep going at a constant speed, and in a straight line, unless something messes with it.
  - Second Law: If you are going to change an object's speed and direction, the amount of force that you apply and the mass of the object determine just how much effect you have — how much you will accelerate the object. The Second Law also states that a particular amount of force used on a more massive object will produce less acceleration.
  - Third Law: If you apply a certain amount of force to an object, the object will apply that same amount of force back to you. That's why, when you throw a ball, you feel pressure on your hand. Same thing when you catch

a ball. And like the Second Law says, the faster the ball goes, or the heavier the ball is, the more force is required to get it moving, or to stop it.

- So, put all that together with an apple that has just cut its connection to the branch, and the First Law of Motion says it would just hang there unless something messes with it. It fell, so something messed with it. Newton made the obvious conclusion that what was messing with the apple was gravity.
- Newton's theory of gravity says that all objects with mass attract each other. They attract each other in proportion to multiplying the value of their masses together, but that attraction gets reduced with distance.
- According to the Third Law of Motion, this gravitational attraction acts equally on both objects. According to the Second Law of Motion, if one of the objects is a lot more massive than the other, this force won't accelerate the more massive object much — it will accelerate the less massive object more dramatically.

Newton's Laws of Motion and theory of gravity were an incredible leap forward. These revelations, and the revelations of other scientists from the Scientific Revolution, resulted in industrial, chemical, and technological advances that reshaped modern human life.

Whether you approve of the modern world or not, it cannot be denied that it was the Scientific Revolution that broke us free from the dogma and tradition of the Middle Ages and brought us to the internet age, to conquering smallpox, to launching probes to explore the solar system, to putting humans (however briefly) on the Moon.

Activity

Newton's Laws and Gravity

We are going to play with how Newton's 2nd and 3rd laws of motion interact with the force of gravity. "Gravity" here is briefly simulated by a stretched bungee cord.

Supplies for this activity:

- A bungee cord (or spring) that is maybe two feet long. You can use a longer bungee cord and just wrap up the ends.
- Safety glasses would be a really great idea, just in case a bungee cord comes loose and flings its metal hook at your face.
- Two objects of vastly different weights. For instance, a cast iron pot and a pencil. Or a box full of books and a shoe. Choose objects that are durable, but that will slide easily.
- Some tape to secure the bungee cord to the two objects. Blue tape (aka painter's tape) is great because it cleans up without leaving a sticky mess.
- A yardstick or tape measure would also be great, but not required.

You are going to demonstrate a couple of Newton's laws of motion. **Whenever one object exerts a force on another object, the second object exerts an equal and opposite force on the first**. That's the Third Law of Motion, and that's where the bungee cord comes in. **The acceleration of an object depends on the mass of the object and the amount of force applied**. That's the Second Law of Motion, and that's why you've got objects of different masses, plus a tape measure.

Find a smooth surface, preferably a smooth area on the floor, or lay down a big piece of cardboard.

Lay down the measuring tape or yardstick (or use some blue tape) to mark where you will be stretching the ends of the bungee cord to — where your objects will start at.

Use some blue tape to secure the ends of the bungee cord to the two objects, if needed. You might need to use a lot of wraps of tape to keep the bungee cord from coming loose when you stretch the bungee cord.

Stretch the bungee cord. The tension in the bungee cord is applying an equal and opposite force on the two objects. It's pulling the object on the left toward the right, and it's pulling the object on the right toward the left. This is sort of like a gravitational attraction, except that the tension of the bungee cord is going to run out pretty quickly when you let go, and gravity never runs out.

Let go. I'm betting that the lighter object moved farther than the heavier object. And maybe the heavier object didn't move at all.

In this same way, when an apple breaks loose from its tree branch, you see the apple fall down to the Earth, but you don't see the Earth fall up to the apple! An equal and opposite force has been applied to both, but the less massive object is the one that moves more.

In this video, first I review Isaac Newton's Laws of Motion, then I imagine what he might have thought about when he watched the apple fall, then I demonstrate this chapter's activity. Enjoy!

https://www.whatacuriousworld.com/chapter-6/#Video06

Links

Biography

27. "Isaac Newton | Biography, Facts, Discoveries, Laws ...." https://www.britannica.com/biography/Isaac-Newton.

Deeper Dive

28. "Scientific Revolution - Wikipedia." https://en.wikipedia.org/wiki/Scientific_Revolution.

Deeper Dive

29. "Newton's cannonball - Wikipedia." https://en.wikipedia.org/wiki/Newton%27s_cannonball.

Deeper Dive

30. "May 25, 1945: Sci-Fi Author Predicts Future by Inventing It ...." https://www.wired.com/2011/05/0525arthur-c-clarke-proposes-geostationary-satellites/.

Great Video

31. "Galileo's Measure Of Gravity Explained By Jim Al-Khalili - YouTube." 9 Nov. 2019, https://www.youtube.com/watch?v=ZBr8Q2ROX9s.

Physics-Speak

32. "Newton's Laws of Motion | Glenn Research Center | NASA." 25 May. 2021, https://www1.grc.nasa.gov/beginners-guide-to-aeronautics/newtons-laws-of-motion/.

Physics-Speak

33. "gravity - Newton's law of gravity | Britannica." https://www.britannica.com/science/gravity-physics/Newtons-law-of-gravity.

7. So Close, and Yet — So Far

"It doesn't matter how beautiful your theory is,
it doesn't matter how smart you are.
If it doesn't agree with experiment, it's wrong."
~ Richard Feynman

By the time the year 1800 rolled around, Newton's "Laws of Nature" had defined physics for a century. While Newtonian physics would continue to rule for another century to come, experimental results started cropping up in the 1800s that challenged the axioms, the assumptions: the very roots of Newtonian physics.

So what exactly undermined Newtonian physics, you ask?

Well, OK, you didn't ask but I'm pretending that you did. Come along with me for a look at what experiments revealed in the 1800s. It gets "interesting." Experimental results from the 1800s, and the efforts to understand them, have produced some of the strangest, most accurate, and (in my humble opinion) most beautiful, descriptions of reality to date.

These experiments revealed that two sets of assumptions used in Newtonian physics were flawed.

One set of assumptions was that space and time were not flexible. Newtonian physics assumes that space is the static background of the universe, the fixed stage upon which the objects and actions of the world play out. It also assumes that time marches on without variation. When one second ticks by here, that same second ticks by everywhere in the universe. In the chapters that follow we will explore the experimental results that challenged the classical views held by Newton about the nature of space and time.

Einstein developed his theory of **special relativity** to explain these experimental results. To bring theory into agreement with observations made in the 1800s, special relativity redefines our concepts of space and time. This would not interest us here, except that special relativity led Einstein inexorably to his theory of **general relativity**, which redefined gravity. And gravity, of course, is what this little book is all about.

I'll give you the $1.00 tour on this Einstein guy's personal background in Chapter 9.

Newton's theory of gravity produced amazingly accurate results, almost all of the time. Almost. Einstein's theory of gravity, presented in general relativity, absolutely nailed it.

In every test that we have been capable of making, Einstein's theory exactly matches how gravity really works.

So to sum up: for our normal everyday lives, Newtonian physics is plenty good enough — as long as you don't look too closely at light, or at the orbit of the planet Mercury. When folks looked too closely at light, and at a curious distortion of Mercury's orbit, Newtonian physics started looking more and more like a very good approximation of how the universe works, but not the "Laws of Nature." But don't sell Newton short.

"Any man whose errors take ten years to correct is quite a man."
~ J. Robert Oppenheimer

It took a century to start noticing flaws in Newton's assumptions. It took another century to correct them. Isaac Newton was "quite a man" indeed. And unless you work at a particle accelerator, study cosmic rays, study the orbit of the planet Mercury, or study the light and motion of distant stars and galaxies, Newtonian physics is probably close enough to match your facts — your "observed phenomena."

In the next chapter we will explore the revelations of the 1800s that led to Einstein's theories of special and general relativity. It had to do with figuring out what light is, and how to measure how fast it goes. Those observations showed that our universe gets a little odd when you approach the speed of light. As Morpheus says in the original Matrix movie "This will feel a little weird."

This chapter is short and sweet, so no review, activity, or links are required… but I *will* repeat the most excellent quote that I started the chapter with, and request that you give it some serious thought.

"It doesn't matter how beautiful your theory is,
it doesn't matter how smart you are.
If it doesn't agree with experiment, it's wrong."
~ Richard Feynman

P.S. The other set of Newtonian assumptions that got blown apart by experimental results in the 1800s had to do with how nature operates at the smallest scales of existence. Those experiments resulted in the theory of quantum mechanics. But quantum mechanics doesn't *directly* affect gravity. On the other hand quantum mechanics *does* raise questions about our best theory of gravity. And *I* think that quantum mechanics is just too important to let slide by, so I have added a couple of appendixes to the end of this book to cover that subject. If you are curious about this set of experiments, and the theory that was put together to explain those results, the appendixes are there for you.

8. Secrets Hiding in the Light

"Your assumptions are your windows on the world.
Scrub them off every once in a while,
or the light won't come in."
~ Isaac Asimov

This chapter covers the experimental results that led up to the creation of special relativity. I'm showing you the progression of experimental results that led to the theory of special relativity, so that you can understand *why* it was created.

The next chapter delves into special relativity itself. Why bother you with this special relativity in a book about gravity? Because Einstein's special relativity laid the groundwork for his incredibly accurate theory of gravity: general relativity. I could have just skipped over special relativity and gone straight to general relativity, but that would be like saying "here's Einstein's gravity, take it or leave it." I'd much rather show you the chain of discoveries that forced the creation of the theories of special and general relativity. I'd much rather give you a glimpse of just how science happens.

It all started with the study of electricity, magnetism, and light. Let's take a brief look at how our understanding of light, and electromagnetic fields in general, unfolded.

Invisible Forces

Michael Faraday[34] was born in 1791, in a village just south of London. He was largely self-educated. When he was 14 he apprenticed to a bookbinder/bookseller, which gave him access to a wide range of material. He was particularly fond of books on the improvement of one's mental faculties and also books on chemistry.

By the end of Faraday's apprenticeship with the bookmonger, a patron named William Dance had been providing Faraday with tickets to chemistry lectures. Faraday was particularly impressed by the lectures by Sir Humphry Davy of the Royal Institution and the Royal Society. Faraday presented Davy with a 300-page book that Faraday had written, based on his notes from those lectures. This so impressed Davy that he took on Faraday as a personal assistant, and had him posted in the Royal Institution as Chemical Assistant.

As a chemist, Faraday synthesized new compounds, formulated a heavier glass for optical purposes, studied the diffusion and liquefaction of gasses, invented a precursor to the chemist's ever-present Bunsen burner, and experimented with what would these days be called "nanoparticles" of gold, suspended in solution.

But what Faraday is truly known for, and what started a chain of experimentation that eventually undermined Newtonian physics, was the study of electricity and magnetism. The study of invisible forces.

In the early 1800s, he embarked on a series of experiments that suggested that electricity and magnetism were but two sides of the same coin. He also advanced the idea that the various forms of known electricity (fish with an electric charge, amber rubbed with fur, lightning, and so forth) were really different expressions of a single "electric force." The results of Faraday's experiments paved the way for the creation of electroplating, electric motors, electric generators, and by extension everything that runs on electricity. And where would our tablets and cell phones be without electricity?

To power his experiments he used an early type of battery called a voltaic pile.[35] It is said that his voltaic pile was cobbled together from copper halfpennies, disks of zinc, and paper moistened with salt water. This design was invented by the Italian scientist Allesandro Volta (hence the name "voltaic pile") and passed on to Faraday when Faraday and his mentor Sir Humphry Davy visited Volta's laboratory.

In his 1831 experiment, Faraday used his voltaic pile to run an electric current through a wire coiled around a narrow tube. In the first step of this experiment, he would move this narrow electric coil up and down inside a larger tube. The larger tube also had a separate wire coiled around it. Moving the narrow tube's electrified coil within the larger tube caused a new electric current to be induced in the larger tube's wire without any physical contact between the two coils.

The second step in the experiment was that the wire from the larger coil was looped out under a freely suspended magnetic compass needle, and when a current was induced in the larger coil, the compass needle was deflected away from magnetic north. The compass needle was deflected without any physical contact with the wire.

Taken as a whole, this experiment demonstrates that electricity and magnetism are intimately intertwined, or might actually be two different expressions of the same force.

By inducing a current in the larger coil, it strongly supported the idea that the magnetic field created by the electric current in the narrow coil was generating a new electric current in the larger coil. In deflecting the magnetic compass needle, it was obvious that the current was creating a magnetic field. Bits and pieces of this experiment had been pioneered by others, but Faraday put it all together in one simple package.

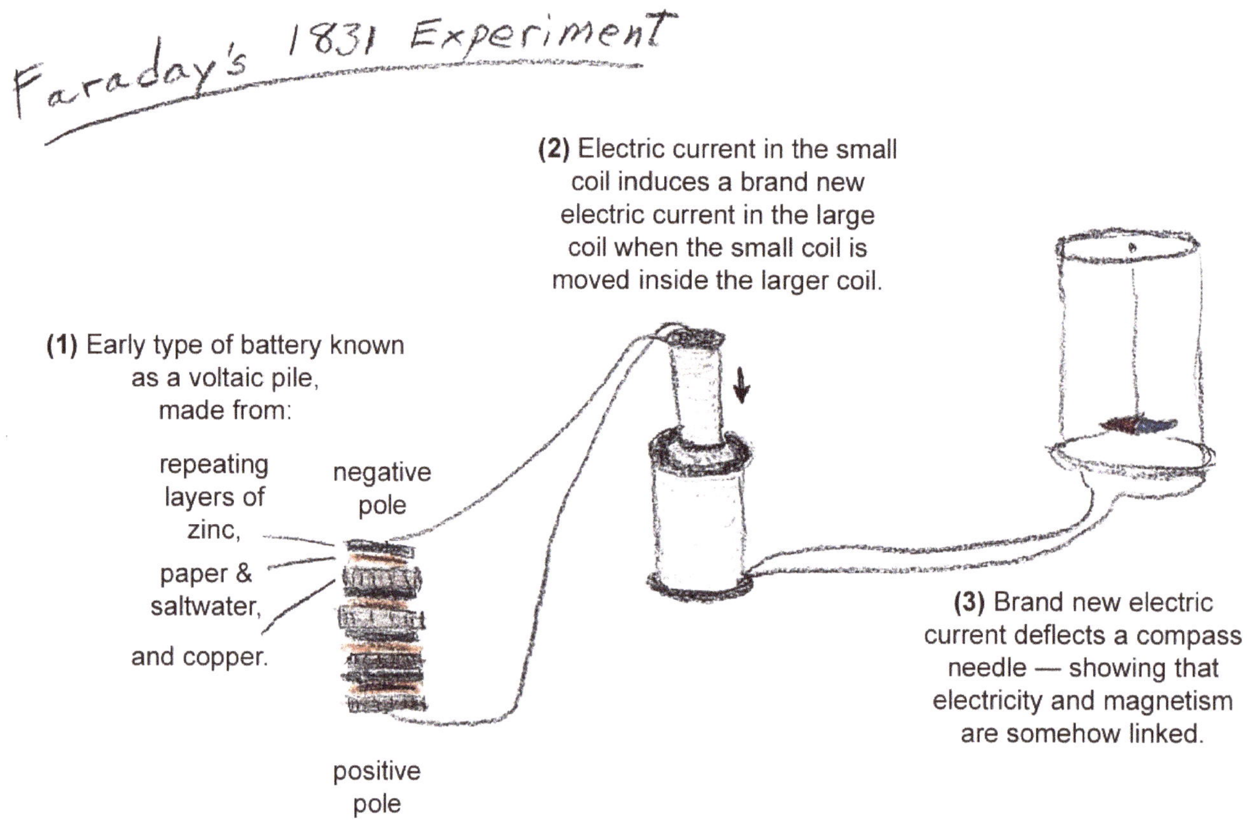

Faraday's 1831 Experiment

(2) Electric current in the small coil induces a brand new electric current in the large coil when the small coil is moved inside the larger coil.

(1) Early type of battery known as a voltaic pile, made from:

repeating layers of zinc,

paper & saltwater,

and copper.

negative pole

positive pole

(3) Brand new electric current deflects a compass needle — showing that electricity and magnetism are somehow linked.

This got Faraday thinking about the nature of magnetic attraction and repulsion. Faraday noted the pattern that iron filings make when sprinkled next to a magnet. The curved lines of the iron filings radiating from the magnetic poles inspired his realization that the magnet was creating an invisible field of influence in the space around itself. This idea of a "field of influence" was later seized upon by Einstein as he struggled to form his relativistic theory of gravity.

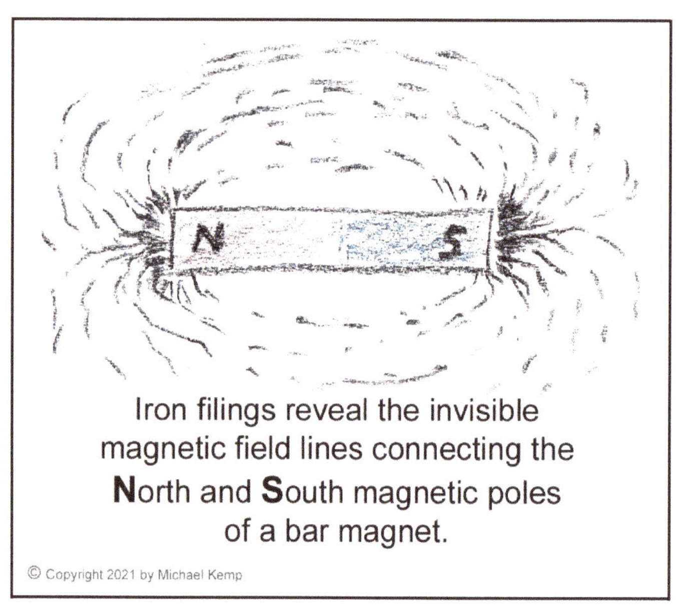

Iron filings reveal the invisible magnetic field lines connecting the **N**orth and **S**outh magnetic poles of a bar magnet.

As a side note, you might ask, "If magnetism is created by an electrical *current*, how can a bar magnet or a lodestone — a bit of metal, or a rock, made of iron and just sitting there — generate *their* magnetic fields?"

In the 1920s a couple of physicists demonstrated that electrons have a tiny magnetic field created by their spin, independent of whether they are in motion or not. Electrons that are bound up in atoms tend to pair up in such a way that their magnetic fields cancel each other out. But not always. Iron is one of those atoms in which the atom as a whole is an incredibly tiny magnet, due to its electrons.

In a typical chunk of iron, the myriad atoms' magnetic fields are randomly oriented, and their magnetic fields cancel each other out. But also in iron, a significant portion of the atoms' individual orientations *can* be persuaded to line up, such that their weak individual magnetic fields become a combined magnetic field that even us big, bumbling humans can notice. When this occurs in nature, the resulting rock is called lodestone. When we force the atoms to all line up, when we manufacture a magnet, we call that a bar magnet, or a permanent magnet.

This first video, from Fermilab, walks through each step from electron spin to permanent magnet.[36] The second video, from PBS Space Time, dives into how this quantum spin isn't what we think of when we say "spin." Fasten your seatbelt.[37]

Electric Light

Faraday had tied together magnetism and electricity, but it took a man named James Clerk Maxwell[38] to create a comprehensive picture of the electromagnetic spectrum.

Maxwell was a Scottish laird of a large estate. His contributions to science spread across many fields, but it was his insight into the nature of the electromagnetic spectrum, published in 1865, that interests us here.

He noted that electromagnetic waves have a frequency and wavelength. Frequency is how fast a wave's peaks and valleys go past a fixed location, given in cycles per second. Wavelength is the distance between the wave's peaks. The higher the frequency, the shorter the wavelength. The lower the frequency, the longer the wavelength.

Maxwell created mathematical formulas that matched the accumulated experimental results concerning magnetism and electricity. Maxwell recognized that visible light itself was a small portion of the electromagnetic spectrum. His equations placed the velocity of electromagnetic waves at what is conveniently called "the speed of light." Maxwell's proposal, that light itself was an electromagnetic wave, tied together areas of study that had formerly been seen as independent of each other.

Maxwell built on the pioneering work of Faraday and others to create the equations that are used to this day for describing and controlling electricity. Maxwell's equations are used for designing everything from the generators that power our electric grid, to the hardware that enables both wired and wireless communications.

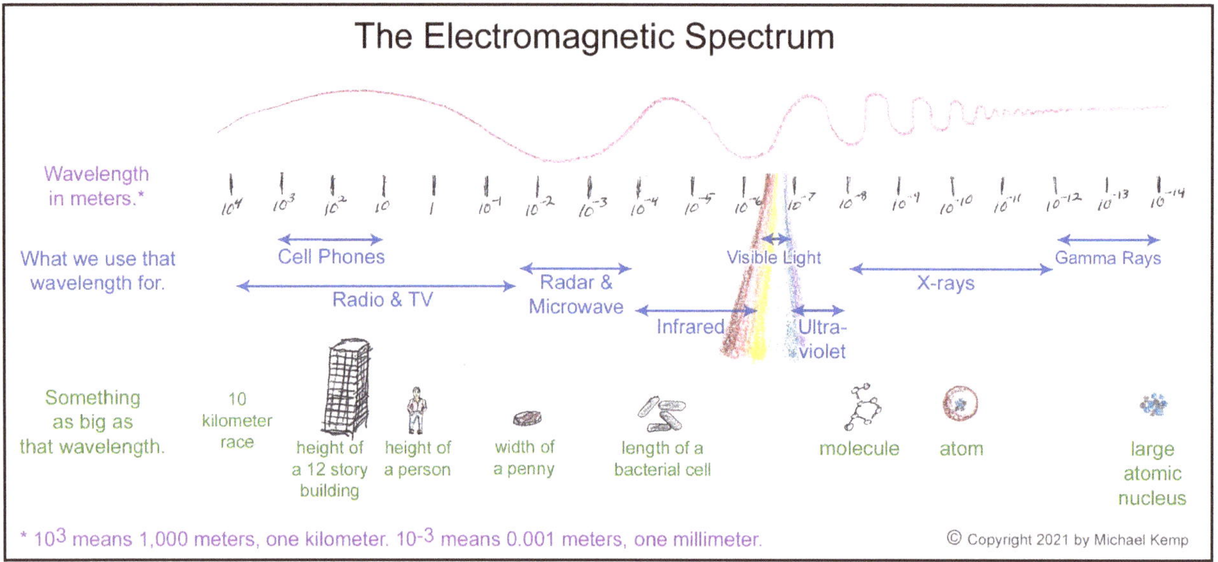

OK, so now we have a basic idea of what light is.

How To Catch a Beam of Light

Does light travel at a particular speed (as Maxwell's equations calculated), or is light instantaneous? Attempts had been made for centuries to answer this question.[39]

To the casual observer, light appears to travel instantaneously. When you flick on a light switch, the whole room is instantly illuminated. When you move a flashlight in the dark, the beam of light appears to move all at once, as if it were a rigid lightsaber (without cutting stuff in half, of course). So you can imagine what a daunting project it was to attempt to measure the speed at which light travels.

The speed of light has been a bone of contention in Western thought since the time of the ancient Greeks. Our old friend Aristotle believed that light was instantaneous, and argued the point with a guy named Empedocles — who maintained that light moves at some unknown speed. Empedocles turned out to be right. On the other hand, Empedocles thought that we see things by light shining *out* of our eyes rather than *into* them. Plus one, minus one for good old Empedocles.

Moving right along, in the early 1600s, Galileo tried to actually *measure* the speed of light. The only conclusion that he could draw from his experiment was that light traveled at least an order of magnitude faster than sound. In other words, at least two miles per second. So Galileo gets a participation award, but no medal.

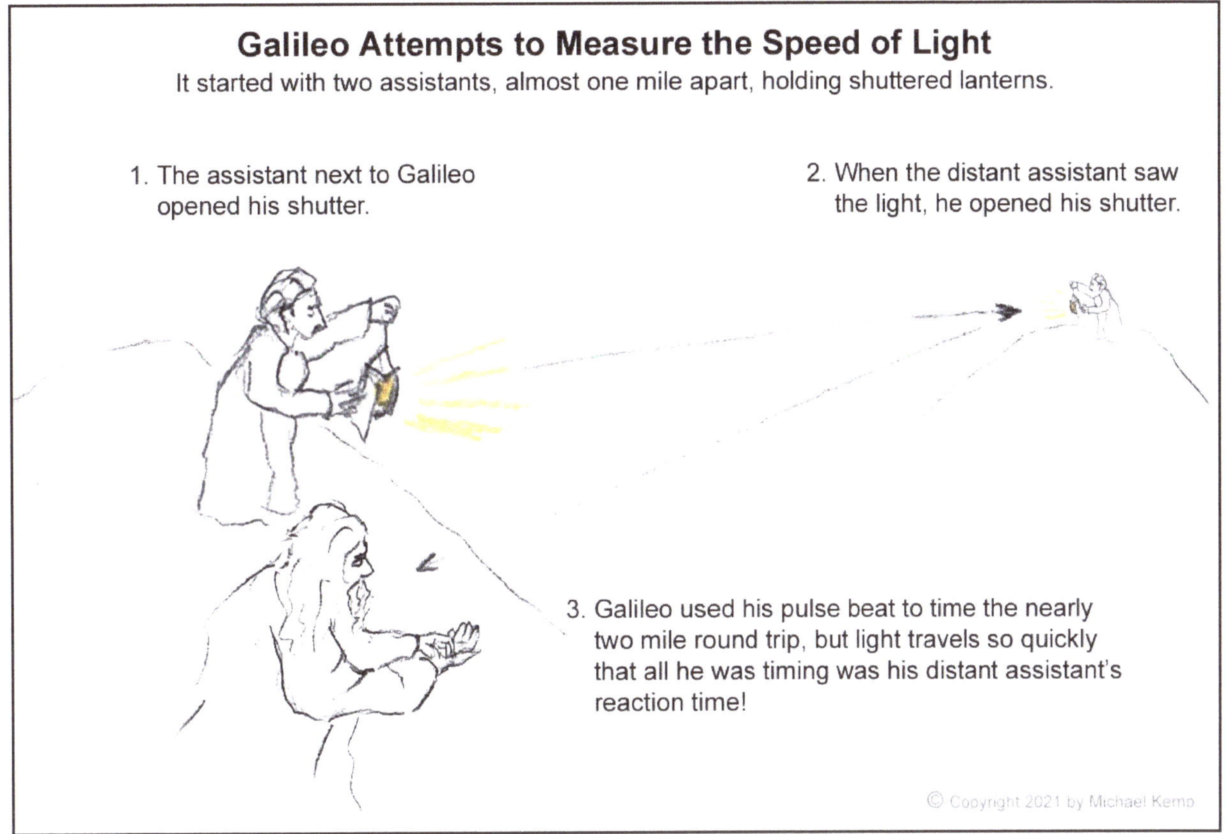

In 1676, the Danish astronomer Ole Römer tried using celestial bodies in our solar system as timepieces and came up with a speed variously reported as between 120,000 and 140,000 miles per second.[40] Now we're at least getting into the right ballpark. I'll give Römer a bronze medal for that.

Half a century later, in 1728, the English astronomer/priest James Bradley[41] looked farther out. In attempting to use parallax to determine the distance to a star in the constellation Draco, the various observations didn't quite add up… unless light itself had traveled at some speed, rather than being instantaneous. Bradley came up with 185,000 miles per second. Getting close, very close! That's worth a silver medal.

A century later, in 1850, two Frenchmen, Hippolyte Fizeau[42] and Leon Foucault[43] executed Earthbound experiments that were very much like Galileo's failed attempt. That is, they were similar to Galileo's in concept, but these experiments relied upon mechanical devices, rather than human reaction times, to flicker light on and off. For many years these two were friends and collaborators, but they seem to have had a falling out at the time of their attempt to measure the speed of light, and each developed their own experimental devices.[44] Like Bradley before them, they each came to within 1,000 miles per second of what we accept today as the speed of light in a vacuum. That ain't bad. More silver medals all around.

Born in 1852, Albert A. Michelson[45] was destined to *really* shake things up. The son of Prussian immigrants, he grew up in California. In school, he demonstrated a gift for science. And in particular, he loved light. He gained the resources to advance his study of light through his career at the U.S. Naval Academy. He won an appointment to the Academy in 1869, and after graduating from the academy and then spending a couple of years at sea, he settled back into the academic world. He never relinquished his passion for hands-on experiments.

First, he tackled the problem of getting a more accurate empirical result for the speed of light. In 1879 he used optical equipment of his own enhanced design to measure the speed of light.[46]

Michelson's 1879 experiment was essentially an upgraded version of Foucault's device. Michelson set his experiment up in a building at the U.S. Naval Academy in Annapolis, Maryland. Light entered the building through a small slit in one corner of the building. At the opposite end of the building, this beam of light hit a rotating mirror, and pulses of light from this rotating mirror escaped a hole in another corner of the building, traveling to a distant stationary mirror that reflected the light back into the building (through the same hole that it had come out of) and back to the rotating mirror. Adjustments to the speed of the rotating mirror allowed the re-reflected pulse of light to be directed back to the first corner — projecting the flashes of triple-reflected light on the wall next to the slit where the light originally entered the building. Once the triple-reflected light was resolved in this way, the speed of light could be determined by the geometry of the experiment and the speed of the rotating mirror. The result was 186,380 miles per second, off by 98 miles per second from the currently accepted speed of 186,282 miles per second. What's 98 miles per second among friends? We have a gold medal winner!

Æthereal Dreams

In the early 1800s it had been demonstrated beyond any doubt that light travels as a wave. Water waves travel through water, sound waves travel through the air. It only makes sense that, because light had been shown to travel as a wave, it must have a medium to travel through. This hypothetical medium was nicknamed the **luminiferous æther**. This æther was believed to permeate the entire universe: otherwise, how could light from the stars travel to the Earth?

Obsessed with the study of light, and encouraged by his success at measuring the speed of light, Michelson teamed up with American chemist Edward Morley[47] to investigate the nature of the luminiferous æther. A chemist might seem like an odd choice for a collaborator when one is measuring æther, but Morley had earned the reputation of being an exceptionally gifted experimenter, praised for measuring the atomic weight of oxygen, as well as calculating the densities of a variety of gasses.

Given that the æther was believed to pervade the entire universe; given that the Earth both rotates on its axis and orbits the Sun; and that the Sun follows an orbit around our Milky Way galaxy: all these movements add up to a very complex and compound motion of the surface of the Earth. It would be ridiculous to assume that the universal luminiferous æther would match all of those motions. The only reasonable assumption was that the surface of the Earth must be moving relative to this æther. Michelson was determined to measure the speed and direction of the Earth's motion through the æther.

To understand the Michelson-Morley experiment you need to think about how an expanding wave acts in a medium that is moving relative to the observer. Imagine yourself standing on the bank of a smoothly flowing river. If you drop a stone into the river, right in front of you, ripples from the stone will expand out on the smooth surface of the river. The ripples will expand in circles, radiating out from the center of the splash. But the whole splash-and-ripples pattern will be carried downstream by the flow of the river. For this thought experiment, let's imagine that the flow of the river is faster than the speed at which the ripples are expanding. The downstream edge of the ripple will be expanding even as it is carried downstream by the current. It will travel away from you at the speed of the expanding ripple *plus* the speed of the river. The upstream edge of the ripple will be expanding against the flow of the river. It will be moving away from you at the speed of the flow of the river minus the speed at which the ripples are expanding. In a similar way (but with more angles and geometry involved) the sides of the ripple toward the bank and toward the middle of the river will also have their own speeds relative to where you are standing on the bank. So the speed of different parts of the ripple vary, relative to where you are standing on the river bank.

Michelson and Morley used this idea, that the speed of a wave will be affected by the flow of the medium that it is traveling through, to design and execute an experiment delicate enough to detect the difference in speed between a beam of light traveling against the flow of the æther compared to a beam of light traveling at a 90°angle to the motion of the æther.

They used a clever system that split a single beam of light, then recombined the beam after its split parts had traveled at a 90°angle to each other. The flow of the æther (like the flow of a river) should affect the speed of the two branches of the beam differently. Any tiny difference in the speed of the two branches of the beam of light would be recorded as an interference pattern in the recombined beam of light.

The **Michelson-Morley Experiment** was designed to reveal the speed and direction of Earth's travel through the luminiferous æther.This æther was assumed to exist as a medium for light waves to travel through.

A beam of light was split down two paths. If the light traveling on one of those paths had to fight the flow of the æther, then that beam of light would travel slower than the beam of light traveling on the other path. This would cause an *interference pattern* when the two beams of light were recombined at the detector.

The entire experiment was floated on a pool of mercury to dampen any vibrations from the environment, and also so that the entire experiment could be rotated to make one or the other of the split beams of light "fight the flow" of the æther and arrive at the detector late.

(1) The light source emits a beam of light (red arrows), which is splilt into two beams by the half silvered mirror.

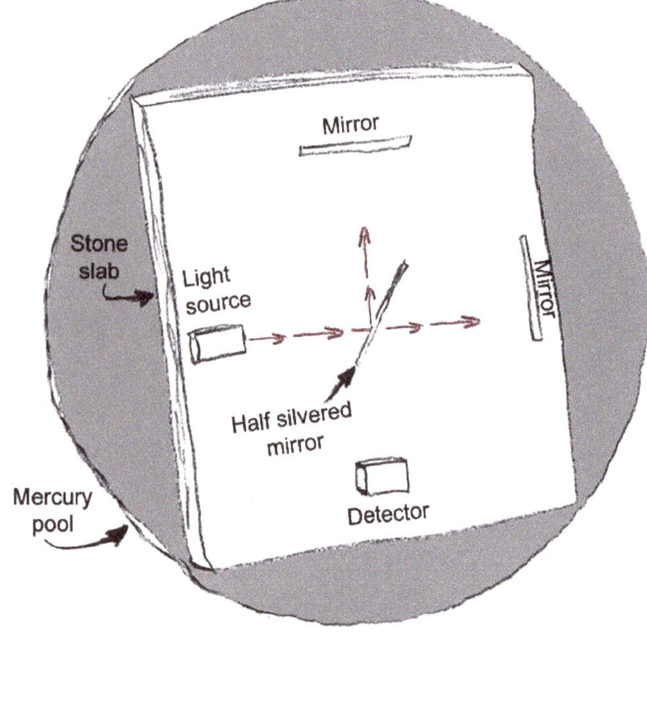

(2) The split beams of light travel at 90° to each other. Then they are reflected off of two more mirrors back to the center of the experiment.

(3) The two beams are recombined and sent to the detector, where any difference in their travel time would be revealed.

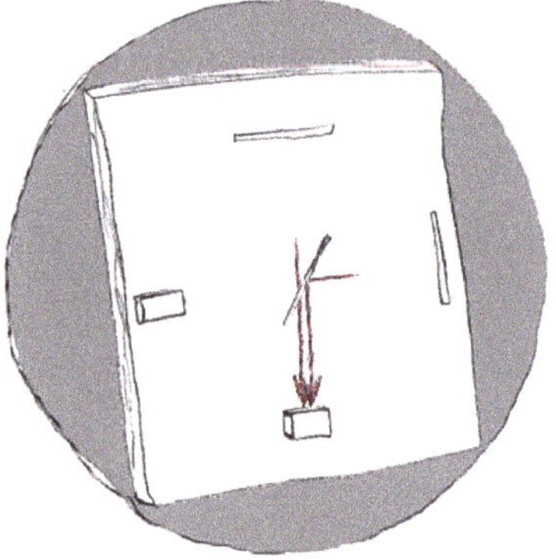

And the experimental result? They could find no difference in the speed of the split beams of light, regardless of the orientation of the experiment. The Michelson-Morley experiment,[48] conducted between April and July of 1887, shook the scientific world's faith in the idea of a universal luminiferous æther.

The results of the experiment showed that light travels at a constant speed no matter in which direction it is oriented. Either the æther is completely synchronized with Earth's compound motions, or there is no æther. Since it would be ludicrous to assume that the universal luminiferous æther is anchored to the compound motions of the Earth, the experiment seemed to disprove the existence of the æther.

More distressing than the possibility that there was no æther — and that light propagates as a wave without any medium — was the revelation that light moves at a constant speed regardless of the motion of the light source relative to the rest of the universe.

Time, Distance, and Mass Get Squishy

Dutch physicist Hendrik Lorentz[49] was not ready to give up on the æther. He put his considerable brain power into translating the experimental results of the Michelson-Morley experiment into a mathematical framework where time and distance were not rigid. That is, he came up with a set of equations that would explain how light traveling in different directions through the flowing luminiferous æther could appear to the observer to be unaffected by the speed or direction of the æther's flow. His equations describe time and distance as elastic, not fixed.

The **Lorentz transformations**[50] are a set of equations that specify how the observed length of a subject is compressed along its direction of travel, how its observed time slows down, and how its observed mass increases. But these things only become measurable when *the observer and the subject* are moving close to lightspeed *relative to each other.* If the speed at which they are approaching each other, or moving apart from each other, is *not* close to lightspeed, then these effects are too small to notice.

There is a phrase, **time dilation**, that you will run into if you read more about special relativity. It's an odd way of saying "time slows down." Dilation generally means "opening up." According to the Lorentz transformations, when a stationary observer watches an object that is traveling at close to the speed of light, you could say that the object's flow of time "opens up" and time moves more slowly for that object. I don't know why they don't just say "slow time," but physicists love the phrase "time dilation," so I thought I'd better give you a heads up.

This is such a bizarre set of ideas that I will say it again. In order to explain the experimental results, Lorentz proposed that when the observer and the subject are moving relative to each other at a significant percentage of the speed of light, the observed length of the subject will be compressed in the direction of the motion; the observed progress of time for the subject will be slowed, and the observed mass of the subject will be increased. In this way, Lorentz's proposed non-rigid space, time, and mass bridged the gap between experimental results and the idea of a luminiferous æther. It was a hack, intended to show how the luminiferous æther could exist, and how it would be impossible to detect the speed of Earth's motions through the æther by looking at the speed of beams of light.

Lorentz's transformations make no mention of the luminiferous æther, they just provide a mechanism for it to remain undetected. On the surface, that seems rather silly, to want something so much that when an experiment fails to reveal it, you devise a scheme whereby it can remain hidden. It's like saying "You can't see my friend, the unicorn, because my unicorn is invisible."

These transformations would not be mentioned in this book on gravity, except for the fact that Albert Einstein latched on to them in creating his special theory of relativity. And special relativity led Einstein inexorably to his theory of general relativity, which is his theory of gravity — his theory of *why stuff falls down* — and that, of course, is precisely what this book is all about.

There are a compelling number of experiments, observations, and real-world applications that confirm Einstein's special theory of relativity. But you won't find Lorentz's unicorn (the luminiferous æther) in any real-world experimental results. And yet, Lorentz's transformations live on in Einstein's theory of **special relativity**.

There are three things I'd like to point out about the Lorentz transformations.

- First, the expanding and shrinking of length, time, and mass are so small as to be undetectable at speeds that we humans experience. They become measurable only when the relative speed between the observer and subject is a significant fraction of the speed of light.

- Second, these calculations are based on a *constant* relative velocity between the observer and subject.

- And third, in the Lorentz transformations, we have another example of answering the question of *how* something occurs, not the question of *why* it occurs.

Well, OK, there is one more little detail. As the relative speed between the observer and the subject approaches the speed of light, the effects of the Lorentz transformations become extreme.

As the relative speed between observer and subject reaches the speed of light, the subject's length along the direction of travel approaches zero, the subject's passage of time slows to a stop, and the subject's mass approaches infinity. This is reality in the observer's frame of reference. This is not an illusion.

For any object that has any amount of mass, for it to approach lightspeed, an infinite amount of force would be required to push the infinite relativistic mass of that object to the final lightspeed velocity. By the equations of the Lorentz transformations, nothing with mass can reach the speed of light.

The Lorentz Transfomations — Illustrated

Pat is observing two boxes, each of which has a clock synchronized with Pat's clock, each box also contains a bowling ball. Each box starts out one foot high and two feet long. The bowling ball weighs 10 pounds.

One of these boxes sits next to Pat. It's speed relative to Pat is **zero** miles per second. The clock in that box ticks at the same speed as Pat's clock. The bowling ball weighs 10 pounds. The length of the box is two feet.

The other box is speeding by Pat at **86.6% of the speed of light**. In Pat's frame of reference, the box's length (along the direction of travel) is compressed to one foot, the clock in the box ticks at half the speed of Pat's clock, and the bowling ball weighs 20 pounds.

This is not an illusion, at least not in the way that we usually think of illusions. In Pat's frame of reference, this is reality. This is how our universe actually works.

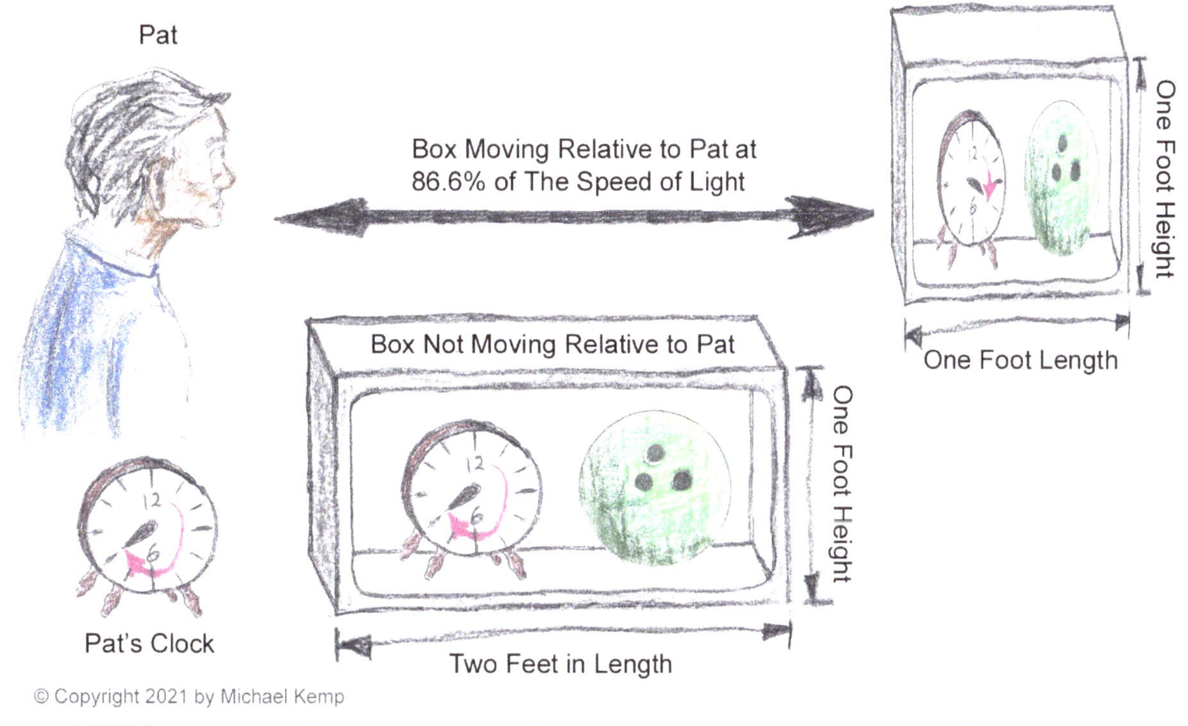

Compressing and expanding distance, time, and mass — is this just how things *appear*, or is it *real*? It is easiest to think of all this as some sort of illusion, some trick of how things appear. Einstein turned the tables, saying that this is not an illusion, this is real. To Einstein, the Lorentz transformations were not a matter of appearances, they were descriptions of how the universe actually operates. Einstein abandoned Newtonian ideas of fixed lengths, universal time, and conservation of mass and used the Lorentz transformations to develop his theory of special relativity.

If we'd never studied light, we could have kept our Newtonian physics unchallenged. But, we are a curious breed. And I mean that in more ways than one.

Review

Michael Faraday demonstrated beyond any doubt that electrical current and magnetism were intimately linked, and possibly just different expressions of the same force. His description of magnetism as an invisible field of force existing in space was critical in future studies of electromagnetism and even influenced Einstein's theory of gravity.

James Clerk Maxwell took it up a level, describing how electromagnetic waves travel, and how light itself is just a small portion of a vast electromagnetic spectrum. Maxwell's equations laid the foundations for harnessing electricity and manipulating light. All the electrical appliances in your life, your cell phone, and even the internet were made possible by James Clerk Maxwell's equations of electromagnetism.

Stuff really hit the fan when the Michelson-Morley experiment demonstrated that light travels at a fixed speed regardless of what direction it is pointed. This showed that luminiferous æther, which electromagnetic waves were supposed to be traveling through, might not exist. Hendrik Lorentz was so upset by this that he devised a set of equations proposing that distance, time, and mass stretch and compress for an object as it approaches the speed of light. This would render the luminiferous æther undetectable. As we will see in the next chapter, Einstein reinterpreted these Lorentz transformations as being the nature of spacetime itself.

I've been fibbing to you. A little. All this stuff about light traveling at a constant speed in any direction, from any source? Well, that's true. Sort of. What *will* slow down electromagnetic waves in general, and visible light in particular, is the medium that the waves are passing through. When light travels through the vacuum of space it is truly going at "lightspeed." It slows down slightly in Earth's atmosphere. It slows down more when it goes through water, and a little more when it goes through glass, and a *lot* more when it goes through a diamond. So when a light wave hits a boundary between mediums it can change speed. If it hits that boundary at an angle, it bends the direction of the wave. This is what causes light to bend when it goes through a prism, or through the lenses of my eyeglasses.

And the shorter the wavelength, the more light gets bent. This is what causes sunlight to make a rainbow when it hits a mist of water droplets in the sky.

For an explanation of rainbows, mirages, and the green flash, follow this link.[51] For a deeper dive into how Maxwell's equations explain why light bends when it hits a different medium, try this link.[52]

But that's all way too far afield for a book on gravity. I just wanted you to know that it's not really that "light always travels at the same speed." It's more like "light always travels at the same speed in the vacuum of space." If you want to get *really* picky, you could say that "light always travels at the same speed, it's just that in a medium like glass, visible light has to spend time interacting with the atoms in the glass. So while an individual photon still travels at lightspeed inside the glass, those interactions give the appearance of slowing down the light."

Activity

The Experiment That Revealed the Unseen

Faraday's 1831 experiment demonstrated the union of electric and magnetic forces. These unseen forces have been manipulated in order to build our modern world. If you have a couple of oddball supplies handy, you can repeat his experiment in your own home. Let's do this!

Supplies for this activity:

- The core of a paper towel roll.
- Painter's tape (aka blue tape or masking tape).
- A plastic compass. Your local sporting goods or outdoor store should have this. Or you can search Amazon for "boy scout compass."
- At least 40 feet of thin insulated wire. Your local hardware store might carry this, or search Amazon for "24 gauge wire." Make sure it is *insulated* wire.
- A 9 volt battery.
- A pair of scissors for cutting the paper towel core.
- A pair of wire cutters for, well, cutting the wire.

Look back at the illustration of *Faraday's 1831 Experiment* near the beginning of this chapter. The first thing we are going to build is those two cylinders with their wire coil circuits.

Cut your paper towel core into three roughly equal lengths. Keep the longest and shortest length sections, and toss the extra section in the recycle bin. If they're all exactly equal in length, just cut an inch off of one of them to make a shorter section.

You now have 2 sections of paper towel core, one slightly shorter than the other.

Take the longer section and cut it lengthwise down one side, end-to-end. Overlap the two sides of the cut ½" and tape them together. This gives you a slightly narrower section of the paper towel core. Make sure that this section of paper towel core fits easily inside the other, slightly shorter section of paper towel core — with room to spare for a layer of wire wrapping. Let's call these two sections of paper towel core the narrow core and the wide core.

Now for the wire wrap.

Leave yourself a foot of free wire before you tape the wire down at one end of the narrow core. Wrap the wire around and around the narrow core, laying each new circle of wire down right next to the last circle of wire, so that you "coat" the narrow core in a single layer of wire in a long spiral of wire wrap. When you get to the other end of the narrow core, tape that end of the wire down to the lip of the narrow core. Cut about a foot-and-a-half of extra wire. Tape it down to the end of the narrow core, then feed that extra wire through the center of the narrow core.

Now do the same thing with the wide core. Except that, when you get the wrap done you do not feed the extra foot-and-a-half of wire back through. Instead, cut ½" of insulation off of both free ends of this wire and twist the bare wires together. This will be the loop of wire that runs under the compass.

Lay the wide core's loop of wire down on the table and place your compass on top of a section of this wire loop such that the compass needle is *not* pointing in the same direction as the wire.

WARNING: You are about to short-circuit the battery. For brief periods of time this is no problem. DO NOT LEAVE THE WIRES ATTACHED TO THE BATTERY FOR AN EXTENDED LENGTH OF TIME. IF THE BATTERY GETS HOT, IMMEDIATELY DISCONNECT THE WIRES.

That said, no problem. If you can't comply with this warning, don't do this activity.

Cut a ½" of insulation off of the loose wire ends of the narrow core. Tape one of these wire ends to each contact on your 9 volt battery.

You are now ready to demonstrate how electrical current and magnetism are just two manifestations of the same force! When you move the narrow core up and down inside the wide core, does the compass needle move?

If not, check out my video to see if I'm doing something differently:
https://www.whatacuriousworld.com/chapter-8/#Video08

Links

Biography
34. "History - Michael Faraday - BBC."
http://www.bbc.com/history/historic_figures/faraday_michael.shtml.

Deeper Dive
35. "Voltaic pile - Wikipedia."
https://en.wikipedia.org/wiki/Voltaic_pile.

Great Video
36. "How Do Magnets Work? - Fermilab."
https://www.youtube.com/watch?v=s7ndBIL402Q.

Deeper Dive
37. "Electrons DO NOT Spin - PBS Space Time."
...that is - they don't spin the same way a top or a wheel spins.
https://www.youtube.com/watch?v=pWlk1gLkF2Y.

Biography
38. "James Clerk Maxwell | Biography & Facts | Britannica." 1 Nov. 2020,
https://www.britannica.com/biography/James-Clerk-Maxwell.

Deeper Dive/Great Video
39. How did we measure the fastest speed there is?
https://www.youtube.com/watch?v=V7PU1WN9jWY

Deeper Dive
40. "Rømer's determination of the speed of light - Wikipedia."
https://en.wikipedia.org/wiki/R%C3%B8mer%27s_determination_of_the_speed_of_light.

Biography
41. "James Bradley | English astronomer | Britannica."
https://www.britannica.com/biography/James-Bradley.

Biography
42. "Hippolyte Fizeau - Wikipedia."
https://en.wikipedia.org/wiki/Hippolyte_Fizeau.

Biography
43. "Léon Foucault - Wikipedia."
https://en.wikipedia.org/wiki/L%C3%A9on_Foucault.

Deeper Dive
44. "Foucault's measurements of the speed of light - Wikipedia"
https://en.wikipedia.org/wiki/Fizeau%E2%80%93Foucault_apparatus.

Biography
45. "Albert A. Michelson - Wikipedia." https://en.wikipedia.org/wiki/Albert_A._Michelson.

Deeper Dive
46. "ALBERT MICHELSON - - MASTER OF LIGHT" https://www.theattic.space/home-page-blogs/2019/12/13/the-master-of-light

Biography
47. "Edward W. Morley - Wikipedia." https://en.wikipedia.org/wiki/Edward_W._Morley.

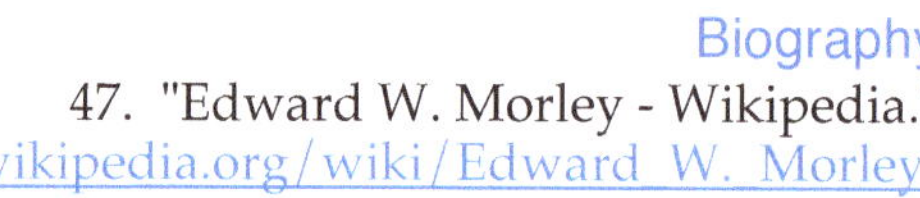

Great Video
48. "Neil deGrasse Tyson explains the Michelson-Morley ...." 21 Jun. 2013, https://www.youtube.com/watch?v=7qJoRNseyLQ.

Biography
49. "Hendrik Antoon Lorentz | Dutch physicist | Britannica." https://www.britannica.com/biography/Hendrik-Antoon-Lorentz.

Great Video/Physics-Speak
50. Lorentz Transformations | minutephysics | YouTube. https://youtu.be/Rh0pYtQG5wI.

Deeper Dive
51. "The formation of rainbows, mirages, and the green flash."

http://www.atmo.arizona.edu/students/courselinks/spring13/atmo170a1s1/1S1P_stuff/atmos_optical_phenomena/optical_phenomena.html.

Great Video/Physics-Speak
52. "Why does light bend when it enters glass? - YouTube." 1 May. 2019, https://www.youtube.com/watch?v=NLmpNM0sgYk.

9. Squeezing and Stretching Reality (Special Relativity)

"Light in empty space is always propagated
with a determinate speed V,
which is independent of the
state of the emitting body."
~ Albert Einstein

Author's note:
The speed of light, which Einstein calls "V" in the above quote, is represented these days with a lowercase "c" — as in the formula $E=mc^2$. This lowercase "c" stands for "celeritas" which is the Latin word for "speed."

The name "Einstein" has become a synonym for "genius," but who was this character, anyway?

Albert Einstein was born in the city of Ulm in the Kingdom of Württemberg in the German Empire in 1879. As a child, he attended Catholic and Prussian-styled schools. In 1894 his father's electrical equipment manufacturing business went belly-up, and his parents left Albert in a boarding house in Munich to finish his schooling while they went to Italy to make a fresh start.

Sick of his school's rote-learning methods, Albert soon dropped out and joined his parents in Italy. To get his education restarted, he took the rigorous entrance exam for the Swiss Federal Polytechnic School. He did great in the math and physics portions of the test, but not so great in chemistry, biology, and French — so he was accepted but had to take make-up classes on the side.

To say that he did great in math and physics is a bit of an understatement. When he was 12, his tutor gave him a geometry textbook which Albert blew through, and then went on to teach himself higher mathematics, leaving his tutor to comment that "the flight of his mathematical genius was so high I could not follow." So you have to keep in mind that this was a kid who taught himself calculus by age 14.

As far as physics goes, his fascination with the invisible force of electromagnetic fields lasted from his first sight of a magnetic compass to his creation of special relativity based on the behavior of electromagnetic waves to his unsuccessful efforts at the end of his life, trying to combine his theory of general relativity with electromagnetic field equations. And don't get me started on $\mathbf{E=mc^2}$ or his insights into the atomic and subatomic realm. So yeah, the guy had a good head on his shoulders for physics and math.

As far as nationality and politics go, Einstein renounced his German citizenship in 1896, apparently to avoid the German military draft. After a few stateless years, he gained Swiss citizenship. In 1911 he took Austrian citizenship while teaching at the university in Prague. Then in 1914, near the start of World War I, Einstein moved back to Germany to teach at the university in Berlin, even though he had gone on record as opposing the German war machine. He regained his German citizenship in 1917, just before WWI officially ended. Then, while Einstein was visiting the United States in 1933, Hitler and the Nazis swept into power. Not only was Einstein philosophically opposed to this new Reich, but Einstein's family was Jewish, and so the Nazi threat to Jews was a very personal threat to Einstein's family and friends. Once again, Einstein renounced his German citizenship. In 1939, on the eve of World War II, Einstein signed a letter (written by a couple of other physicists) to President Roosevelt, warning that Germany could be working on the creation of an atom bomb, and urging the U.S. to beat them to it. The resulting effort became known as The Manhattan Project. And for his *final* citizenship, in 1940 Albert Einstein became a citizen of the United States, a citizenship which he kept for the rest of his days (Einstein died in 1955).

If all that nation-hopping and wars and nuclear weapons drama doesn't leave your head spinning, I don't know what will!

I can't, in good conscience, ignore the behind-the-scenes contributions of Albert's first wife Mileva Marić Einstein. Born in Serbia in 1875, she got into high school in the last year that women were admitted. Her father got her exemptions to allow her to attend boys-only physics lectures.

In 1896 both she and Albert were admitted to the physics-mathematics section of the Polytechnic Institute in Switzerland. They achieved similar class grades at the Institute, but only Albert was granted a degree.

Before Albert's rise to fame, Mileva and Albert co-wrote scientific articles, but published them under Albert's name. This appeared to be a joint decision, in that having a woman's name on the publication would reduce its credibility. And yet, regarding the special theory of relativity, Albert wrote to Mileva "How happy and proud I will be when the two of us together will have brought our work on relative motion to a victorious conclusion."

But such were the times, that her role was not publicly acknowledged. So for the period from around 1900 to 1914, when Albert and Mileva separated, if I say "Einstein" think "Albert and Mileva Einstein."

So that's a little glimpse into Albert Einstein's life journey, but what's all the fuss about the theory of special relativity?

Special relativity is rooted in the revelation that our universe has a built-in speed. We generally call this speed "the speed of light," or "lightspeed," because the study of light was where this speed was first observed. But it is not the nature of light that determines this speed. This speed is actually built into our universe itself. It is the maximum speed at which one bit of our universe can have an effect on any other bit of our universe. This is the fastest speed at which something that happens *here* can cause an effect *over there*. It is more accurately called spacetime's "speed of causality."

How did this all get figured out? Let's start with the question of whether light always travels at a constant speed.

If you think of space and time in the way that Euclid, Galileo, and Newton thought about them — as absolute, constant, and universal — then the speed of light should vary depending upon the relative speed of the observer to the source of the light. A beam of light created by a source that was moving toward you should come past you at the speed of light *plus* the speed that its source was moving toward you. A beam of light created by a source that was moving away from you should come past you at the speed of light *minus* the speed at which its source was moving away from you.

The Michelson-Morley experiment showed that, where light is concerned, 186,282 miles per second plus the speed of the object that created that light equals 186,282 miles per second, regardless of the speed of the source of the light. It's like saying:

- Light created by a source that is moving towards you at 10 miles per second comes past you at 186,282 miles per second: so 186,282 + 10 = 186,282.

- Light created by a source that is moving towards you at 1,000 miles per second comes past you at 186,282 miles per second: so 186,282 + 1,000 = 186,282.

- Light created by a source that is moving towards you at 100,000 miles per second comes past you at 186,282 miles per second: so 186,282 + 100,000 = 186,282.

I don't know about you, but my childhood math teachers would not approve. It couldn't be right. At least not if you stick with the Newtonian concepts of fixed spatial dimensions and universally consistent time. Something had to give.

In order to explain the constant speed of light, which was demonstrated by the Michelson-Morley experiment, the Lorentz transformations proposed that distance, time, and mass start to change near the speed of light. Distance compresses, time slows down, and mass increases.

Newtonian physics had served, and still serves, stunningly well. Newtonian physics provides a highly accurate description of the world as we generally experience it. It is based on the comforting, dependable worldview of the unchanging, fixed nature of space and time.

In Newtonian physics, time is universal and monolithic. That is to say, when a second passes for me, that same second passes for you, and that same second passes instantaneously all over the universe.

In Newtonian physics, space does not bend or contract in unruly ways. A cubit is always a cubit, a yard is always a yard, and a meter is always a meter. It doesn't matter whether you stand on your head and wiggle your ears, or not. It doesn't matter whether you are standing still and leaning against a tree, or whether you are winging by at nearly the speed of light. A yardstick that you are observing does not get all squished down just because you and it are traveling at different speeds!

In our human scale of existence, and at the speeds with which we humans travel, these axioms — these ground-floor assumptions — hold up flawlessly. Newtonian physics just works. Right up to the point where you start looking a little too closely at light.

Einstein Sets Sail On an Elastic Sea of Spacetime

By 1902, Albert Einstein[53] began working as a patent examiner in the Swiss Patent Office. But the gray matter under his curly black hair was working overtime. Einstein knew that the Newtonian model, while being leaps and bounds more accurate and useful than earlier theories about nature, could not be twisted in any manner that would account for the facts revealed by the experiments of the 1800s.

In 1905, Einstein presented four seminal papers that revolutionized our understanding of the fabric of our universe. One of those papers, titled "On the Electrodynamics of Moving Bodies," was the origin of what we now call **special relativity**.[54]

The best way I can think to say it is, that Einstein unmoored himself from the safe harbor of the fixed, rigid Newtonian/Galilean worldview. He set sail on a mental ocean unconstrained by the inflexible concepts of the past, and set his course guided by the 1800s' bizarre experimental results. In the end, his voyage brought him to a more accurate perception of how the universe functions than any human had ever before achieved.

It's one thing to have a wondrous vision. It is quite something else to present explicit equations and mechanisms that transform that vision into a coherent hypothesis. A hypothesis complete with predictions that can be tested in the real world. For this, Einstein utilized the Lorentz transformations and the geometries of four-dimensional space-time pioneered by the German mathematician Bernhard Riemann[55] and fleshed out by one of Einstein's own teachers, Hermann Minkowski.[56] In Einstein's special relativity, the Lorentz transformations are not a description of *how* light travels, they are a description of how the universe itself functions, and *why* light travels as it does.

Special relativity started a 10-year quest by Einstein that culminated in his theory of gravity, also known as **general relativity**. You could say that the variable nature of space and time described by special relativity in 1905 foreshadowed the totally bent nature of the universe that Einstein presented as general relativity in 1915.

Taken together, special and general relativity shattered the foundations of Newtonian physics. And if you want to get a grip on Einstein's gravity (general relativity), it really helps to wrap your mind around its progenitor: special relativity.

Einstein presented three other papers in 1905, which is known as Einstein's "miracle year."[57] One of them was about mass and energy. This paper provided us with the iconic formula $\mathbf{E=mc^2}$. This was the theoretical foundation for the technology that produced nuclear power plants, atom bombs (which used nuclear fission), and hydrogen bombs (which use nuclear fission as the ignition system for nuclear fusion). Suffice it to say that $\mathbf{E=mc^2}$ has been experimentally verified. For a description of nuclear fission versus nuclear fusion, follow this link.[58]

Another paper that he presented in 1905 was on a curious phenomenon called Brownian motion.[59] When subjected to Einstein's inspection, the seemingly random nature of Brownian movements revealed the particulate nature of "stuff." By "stuff" I mean everything from dirt to water to the air that we breathe and the dust in the air. Some of the smallest of these particles are also called "atoms." Please keep in mind that at the time of this paper, the existence of atoms was still a matter of debate.

The fourth paper that he presented in 1905 was about how only certain wavelengths of light cause electrons to be ejected from a piece of metal. This paper sounds particularly dull, but this paper was one of the foundations of quantum mechanics. Quantum mechanics is bizarre and fantastical. If that intrigues you, there is more detail about quantum mechanics in the appendixes.

Picking Your Relatives

Let's back up a second and clarify what physicists mean when they say **relativity**.

Relative to what? Relative to a chosen frame of reference.

The basic idea of relativity is that the theories of physics must apply, without fail, to any chosen frame of reference.

What is a frame of reference? Your personal frame of reference is how you, and no one else, observe your surroundings. For instance: what, relative to you, is stationary or moving? If you are standing on a sidewalk, how fast is a passing car traveling relative to you? If you are riding in a moving car, how fast is a neighboring car traveling in your personal frame of reference? If a neighboring car is moving at the same speed and direction as your car, then its speed, relative to you, is zero.

The most basic form of relativity only applies to a frame of reference that is either stationary or is traveling at a constant speed and direction. This is called an inertial frame of reference. It is considered "inertial" because, true to Newton's First Law of Motion, the state of that frame of reference will remain constant unless an external and unbalanced force is applied to it — in which case it would no longer be "inertial."

Relatively the Same

For what it's worth, relativity didn't start with Einstein. It started with Galileo.

Galileo made his own declarations about an observer's frame of reference. Those declarations are referred to as Galilean Invariance[60] or Galilean Relativity, or they are sometimes called Newtonian Relativity.

It is helpful to ease into Einstein's special relativity with the Galilean/Newtonian version of relativity. The basic idea of Galilean/Newtonian relativity is that **the rules of physics must be maintained when the observer is sitting still, as well as when the observer is moving at a constant velocity** (a constant speed and direction). In other words, when the observer is in an inertial frame of reference.[61]

Galileo used thought experiments involving a sailing ship. In these thought experiments, he maintained that experimental results would be the same, regardless of whether the ship was at rest in the harbor, or smoothly sailing at a constant velocity.[62]

For instance, a rock dropped from the top of a mast would land at the base of the mast, whether the ship was at rest or smoothly making way on a calm sea. In this case, the ship, the mast, and the rock are taken together to be a "closed system."

When discussing physical observations, the term **closed system** means that nothing outside the set of objects being observed should interfere with them. A scientist will choose a group of objects to study. This is the "system" being observed. In the above example, the ship, the mast, and the rock must not be rocked by waves, blown about by gusts of wind, or struck by lightning. Objects in a closed system must not be manipulated by anything outside of the "system."

Galileo also noted that a man on a ship, in a closed cabin with the windows covered, could perform more delicate experiments such as watching the path of smoke from incense, observing the flight of an insect, or tossing a ball in different directions — and he would be unable to determine from his observations whether the ship was stationary or moving smoothly at a constant velocity.

You can't tell from within a closed system whether you are "sitting still" or "moving at a constant velocity."

That's the Galilean version of relativity.

Einstein expanded the idea of relativity in two dramatic ways. His special relativity

- Has two dueling frames of reference, not just a single frame of reference.
- Utilizes the Lorentz transformations.

Traveling a Straight Stretch of Highway at 60 mph

Let's try to get a feel for how dueling frames of reference look. Imagine that you are sitting in the backseat of a blue SUV that is traveling down a straight stretch of highway at a constant 60 mph. Note that the driver of the SUV does not move relative to you. The driver's relative speed (in your frame of reference) is zero.

As you pass a tree at the side of the road, the tree is moving in your frame of reference at 60 mph. That's right. Relative to you, the tree is *moving away from you* at 60 mph.

A red sports car going 70 mph and traveling in the same direction as your blue SUV passes you. Relative to you, the red sports car is passing by you at a rate of 10 mph.

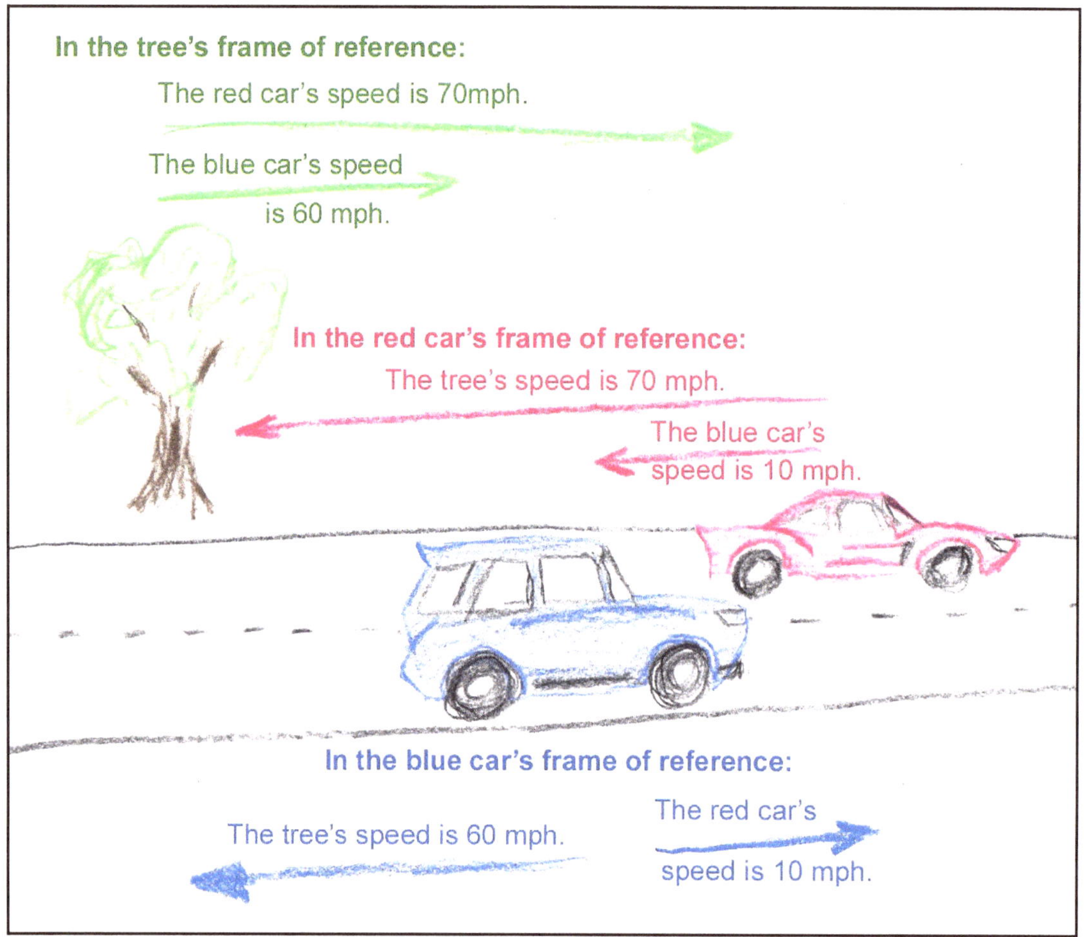

That's what relativity is about: how the objects that you observe move in *your* frame of reference. It's not how an object moves in some sort of "absolute" sense.

In special relativity, you *choose* an inertial frame of reference, and the rules of physics must apply to all of the observations of an object that can be made from that frame of reference. Choose a different frame of reference, and your observations of that very same object may produce different results. The rules of physics must match observed phenomena within each frame of reference, but they are relative to that frame of reference. There is no absolute frame of reference.

Back in the blue SUV, in your frame of reference, the red sports car that is passing you is moving at 10 mph. From the frame of reference of the tree at the roadside, the red sports car is moving at 70 mph. Both of these frames of reference are valid. They are both real. The rules of physics are obeyed within each frame of reference.

In relativity, speed is relative to your frame of reference. It depends on your location. Just like the direction of "down" depends on your location. Are you in the blue SUV, the red sports car, or leaning against the tree?

Leaning Against a Tree

Do you think that you are "stationary," just because you are lounging around, leaning against a tree?

Well, if you live around the 45th latitude as I do, then the rotation of the Earth is about 733 mph. At the North and South Poles, it is essentially zero. At the equator, it is just about 1,037 mph.

That doesn't take into account the speed at which the Earth is orbiting around the Sun (roughly 67,000 mph), or the speed at which the Sun is orbiting the center of the Milky Way galaxy (approximately 448,000 mph). And don't get me started about the motion of the Milky Way itself.

So if you spend an hour leaning on a friendly tree are you *really* stationary? In your frame of reference, why, yes you are. In the Sun's frame of reference, not so much. In the frame of reference of the Milky Way galaxy, oh heck no!

Pick a Frame, Any Frame

Einstein's special relativity is all about how the universe presents itself to you in *your frame of reference*. And about how the universe presents itself to someone else in *their frame of reference*. And about how the real results from those two frames of reference deviate from each other: both real and yet not the same.

A speed relative to you is measured from your frame of reference. The driver of your blue SUV is moving at zero mph relative to you. And this is just as real as the fact that the driver of your blue SUV is moving at 60 mph in the frame of reference of the tree at the roadside.

The *relative* in *relativity* means that observations are made relative to a chosen frame of reference. Pick a frame of reference. They're all real.

Reality Is Relative

As he developed his special theory of relativity, Einstein focused on the motion of the observer and the subject relative to each other, and how the laws of physics must be preserved regardless of any constant difference in velocity between the frame of reference of the observer and the frame of reference of the subject.

To get a feel for how Einstein thought about how reality is experienced by observers in two frames of reference, let's start with a thought experiment using speeds that we are all familiar with.

Einstein seems to have been inordinately fond of thought experiments involving trains. For instance, in part one, chapter three of his book *Relativity: The Special And The General Theory*,[63] Einstein asks us to consider two observers. One is aboard a smoothly running train traveling at a constant velocity. The other observer is standing on a footpath next to the railroad tracks. The observer on the train drops a heavy stone from the train's window. To the observer on the train, the stone falls straight down. To the observer standing on a footpath as the train passes, the stone travels in a smooth curve, combining the forward velocity that it had while on the train with the pull of gravity as it falls.

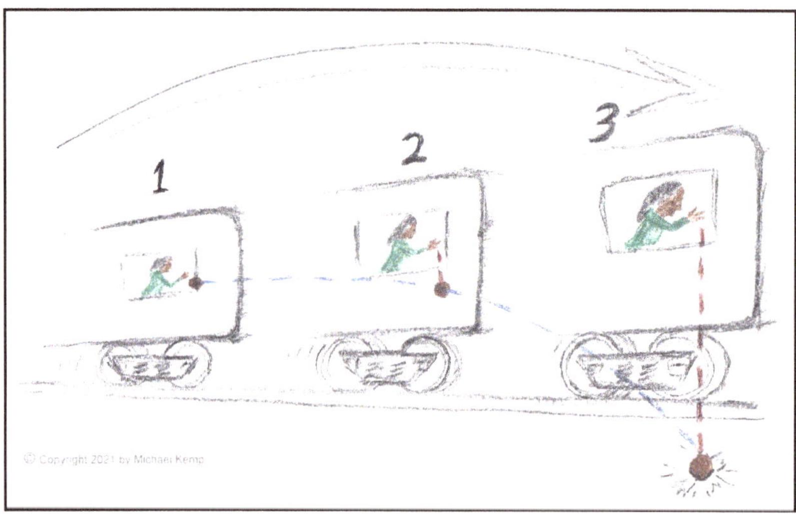

The laws of Newtonian physics are obeyed as far as each of these observers is concerned.

Einstein then poses the question of whether, *"in reality,"* the stone travels through space in a straight line (as seen by the train's passenger, red dashed lines) or in a curve (as seen by the observer on the nearby footpath, blue dashed line)?

Einstein concludes that, in reality, the stone falls *both* in a straight line *and* follows a curve — depending upon which frame of reference you choose. Both are *real*, **relative** to the associated observer.

So not only is beauty in the eye of the beholder, reality itself is relative to the beholder. As with the direction of "down," reality varies depending on the location of the observer.

This thought experiment shows us the different realities experienced by two observers of a single event. Each observer has a valid frame of reference. A single event is experienced differently in each frame of reference. Special relativity is all about relating two frames of reference, with a constant velocity between those two frames of reference. If that relative velocity is near the speed of light, the differences experienced between the two frames of reference get a lot stranger than "Did the stone fall straight down or in a curve?" As you approach the speed of light, the differences take the form of the Lorentz transformations.

Lightspeed Is How the Universe Is Built

Einstein rejected the idea of a luminiferous æther. Einstein took the Lorentz transformations *not* as a description of how light behaves, but as a description of how our universe works at a fundamental level. Einstein reimagined the lightspeed of the Lorentz transformations as the maximum speed at which one bit of our universe can affect any other bit of our universe. This speed is the speed of cause and effect. It is sometimes called spacetime's **speed of causality**.

So, you can see that "lightspeed" is not really a property of light. According to special relativity, it is a fundamental property of our universe. It is the fastest speed at which anything can be transmitted across the four dimensions of spacetime. It is how the universe is built.

Special relativity incorporates the Lorentz transformations to say that as a subject's speed (relative to the observer) nears the speed of light, the subject's length along the direction of travel shrinks, the subject's passage of time slows, and the subject's mass increases — in the observer's frame of reference. In the subject's frame of reference, these effects do not occur. In the subject's frame of reference, length, time, and mass are not altered. Each frame of reference is valid. Each experience is real.

Another thing to remember is that in Einstein's relativity, there is no absolute frame of reference. If you are sitting on a park bench, you may call that "stationary." To someone far out in space, watching the Earth spinning on its axis and orbiting the Sun, you are not stationary — not in their frame of reference.

In the illustration of the blue SUV, the red sports car, and the tree — as long as neither car accelerates, each of those three frames of reference are equally real. Common sense tells us that the tree's frame of reference is "real" and the frames of reference for folks riding in the SUV and sports car are not as "real" as the tree's frame of reference. Einstein says that's wrong. Every inertial frame of reference is real. The tree has no claim to some sort of "absolute" or "better" frame of reference.

The same goes for the example of a stone being dropped from a moving train. In one frame of reference, the stone falls down in a straight line while in the other frame of reference, the stone falls in a curve. Both observations are real. Neither frame of reference has a claim to being better than the other.

Special relativity is interested in two inertial frames of reference that have some large difference in velocity between the two of them. The Lorentz transformations do not come into play unless there is a large difference in velocity between one frame of reference for the observer, and another frame of reference for the object being studied.

An observer on Earth studying an object winging by the Earth at 90% the speed of light will *definitely* see the effects of the Lorentz transformations on that object, from the observer's frame of reference.

If the observer and the object being studied are *both* winging past Earth, side by side, at 90% the speed of light, then they are moving *relative to each other* at zero miles per hour. The Lorentz transformations will have no effect on how that observer sees that object. The issue is not how something moves compared to a common sense *absolute* frame of reference. Special relativity says there is no such thing as an absolute frame of reference. Instead, each and every bit of the universe that is not accelerating has its own valid inertial frame of reference.

As we will see at the end of the chapter, experimental results confirm that this is how the universe works. It's not an illusion.

Time is a Location

In our daily lives we usually think of the space that we move through as having three dimensions. If I am standing out in a field, I know that I can change my location by moving left or right (1st dimension). I can also jump up or fall down (2nd dimension). And I can move forward or backward (3rd dimension).

As Europeans started exploring our planet, they came up with a way to define a location that worked for the entire globe. Latitude tells you how far you are (north or south) from the equator. Longitude tells you how far you are (east or west) from Britain's Royal Observatory in Greenwich, England (yes, the British Empire was that big of a deal). Elevation tells you how far you are (above or below) average sea level. With those three dimensions, you can specify any physical location on, above, or within the Earth. This is three-dimensional space.

Special relativity also binds time into this system of coordinates. Time is no longer considered to be independent of three-dimensional space. The result is a **four-dimensional spacetime**.

I find it easier to come to grips with four-dimensional spacetime if I think of spacetime as a system of coordinates for locating a person, an object, or an event.

An example: If I want to catch the bus, I have to be at the bus stop (latitude, longitude, and elevation) at the time that the bus is due (that's the location in time). So all four dimensions of spacetime are needed to locate my desired event: "catch the bus." Admittedly, in the real world, the bus might be a little early or a little late — so the exact location in time is a little mushy. But the same is true of latitude and longitude. I could wander a few yards one way or another from the bus stop and still be able to catch the bus when it shows up. (On the other hand, even a small mix-up in elevation could have serious consequences. If my elevation were a little too high I would find myself falling, and crashing to the ground. If my elevation were a little too low I would find myself embedded in the concrete sidewalk. Not good.)

As another example, you might think of traveling from Boston to New Orleans by plane as a three-dimensional journey: you go up, you go south, you go west, and you go down. But to describe your trip more fully, you would record it as a four-dimensional journey including the time of departure from Boston and the time of arrival in New Orleans.

A third example: If you want to locate an event in the past you will need to specify all four dimensions of spacetime. For example, the "Alpha" release of the Minecraft game occurred at approximately 59.3° north latitude, 18° east longitude, and 100 feet elevation (aka Stockholm, Sweden), on June 30, 2010. That was the location in spacetime of this particular event.

A location in spacetime in Earth's frame of reference has a latitude, longitude, elevation, and point in time.

If you want to specify something that is not on planet Earth, you would need to use some other system of spatial coordinates rather than latitude, longitude, and elevation. You could still use Earth-relative time — unless some other way of measuring time was more appropriate.

In order to talk about spacetime in a way that is not restricted to the Earth, the four dimensions are generally represented by the variables: x, y, z, t. So if you are standing out in a big field, you could say that the x dimension represents your ability to move left and right, y could represent the up/down dimension, z could represent your ability to move forwards and backward, and t would be the time dimension. Using x, y, z, t allows you to talk about four-dimensional spacetime, regardless of whether you are considering something here on Earth, something far away (like the orbit of the planet Mercury), or something completely abstract (such as a thought experiment).

In our everyday lives, we are already accustomed to thinking of events in four dimensions. If we wish to meet someone in the conference room on the 12th floor (in other words, at a spatial location that could be precisely given by x, y, and z) we also need to specify a time to meet (the fourth dimension t). So you would actually say "Meet me in the conference room on the 12th floor at 2:00 pm today." In doing this you have given a complete four-dimensional location in *spacetime*.

But Einstein's spacetime is not just four-dimensional. In special relativity, this spacetime is not rigid. In the frame of reference of an observer, for a subject that is traveling at close to the speed of light (relative to the observer), the subject's spatial dimension in the direction of travel shrinks and the subject's time dimension slows down (in physics-speak, their time "dilates"). That makes for some messy geometry!

As mentioned above, to get the equations for dealing with the flexible four-dimensional spacetime of special relativity, Einstein turned to the work of his former teacher, Hermann Minkowski. Something of a math prodigy, Minkowski began his university studies at age 15. Born in Poland, he later taught at universities in Germany and Switzerland. Minkowski's fascination with the intersection of geometry and mathematics paved the way for developing the formulas to describe flexible four-dimensional spacetime. Einstein, one of Minkowski's students, was obviously paying attention in class.

I've talked about how Einstein used the work of other scientists and mathematicians, such as Lorentz and Minkowski, to construct his theory of special relativity, but don't get me wrong. I'm not saying that Einstein just cobbled together other people's work. Einstein (with the help of his wife Mileva) imagined a flexible universe, with the nature of distance, time, and mass tied to the frame of reference of each "observer." This was truly revolutionary.

Einstein's insistence that reality itself is different for different observers fundamentally changed modern physics. Whether considering the observations of two observers watching a rock drop from a moving train, or the experience of reality by an observer and a subject (or by two observers) whizzing past each other at near the speed of light… Einstein insisted that the observations made in each frame of reference were real — even when they seemed to contradict each other.

Spacetime's "Speed" of Causality Isn't *Just* a Speed

"... buckle your seatbelt, Dorothy,
'cause Kansas is goin' bye-bye."
~ Cypher (The Matrix)

Our common sense definition of "speed" is how much distance something travels through the ***x, y, z*** spatial dimensions in a given period of the time dimension ***t***. What's the speed of that car? 60 *miles* per 1 *hour*.

Spacetime's speed of causality can do a lot of other things besides making light travel through the ***x, y, z*** spatial dimensions at 186,282 miles per second. Spacetime's speed of causality can change how things move in the time dimension ***t*** as well.

Spacetime's speed of causality is not what we normally mean by a "speed," because it is movement through *all four spacetime dimensions* ***x, y, z, t*** all put together, not just ***x, y, z*** measured against a given period of time ***t***.

I'll give you a physical way to think about it. You can either treat this as a thought experiment, or take this as an extra-credit activity.

Get four empty gallon jugs. Label one jug ***x***, and label the next jug ***y***, the next jug ***z***, and the last jug ***t***. Now fill the ***t*** jug with a gallon of water. If you don't have gallon jugs, you could use any four jugs as long as they are equal volume, and start with the ***t*** jug full.

- The ***x*** jug is movement left-right in your frame of reference.
- The ***y*** jug is movement up-down in your frame of reference.
- The ***z*** jug is movement forward-backward in your frame of reference.
- The ***t*** jug is movement from the present to the future in your frame of reference.
- The gallon of water represents spacetime's speed of causality.

Think of time as a location in spacetime. Unlike the other three dimensions, you can only move forward in time. Other than that little quirk, time is just another dimension in spacetime, another way to specify a location in spacetime.

In this activity, the entire gallon of water is *always* in use. It can be moved to different jugs, but the full amount of spacetime's speed of causality is *always* in play.

We start with the ***t*** jug being full. When the ***t*** jug (time) is holding the full gallon of water (*all* of spacetime's speed of causality), time is running at normal speed. For our day-to-day experience this is what we observe. We observe this for ourselves, for the passengers in a sports car whizzing by, for anything or anybody we observe in our daily lives. As far as we can tell, all of spacetime's speed of causality is being used in the ***t*** jug, moving us from the present into the future at a normal rate of time.

What if we took one drop of water from the ***t*** jug and put it into the ***x*** jug. Now there is one drop of spacetime's speed of causality in our left-right spatial dimension ***x***. If we saw this in our real life, we would be observing something winging past us at 4 times the speed of a rifle bullet. One tiny drop of spacetime's speed of causality moved into the ***x*** dimension means an object going 4 times the speed of a rifle bullet in that spatial dimension. And the ***t*** jug still looks full. As far as a human being could tell, time would still be running at a normal rate for that object. If you put a total of 3 drops into the ***x, y,*** or ***z*** jugs, that would match the Guinness World Record for the fastest any human has ever traveled (the astronauts in the Apollo 10 command module). Even missing 3 drops, the ***t*** jug still *looks* full. Anybody looking at those Apollo 10 astronauts would say that their time sure looks like it's running at a normal rate (their ***t*** jug sure looks like it's full).

But isn't spacetime's speed of causality also lightspeed? Why yes it is.

Pour all of the water (all of spacetime's speed of causality) into the ***y*** jug. Every drop from every other jug. This is how spacetime's speed of causality is allocated for a photon that is coming straight down in your ***y*** dimension. All of its speed of causality is used in the ***y*** jug. None is left in the ***t*** jug. From your frame of reference, that photon speeds down at 186,282 miles per second, and that photon's internal clock is frozen. Its ***t*** jug is empty.

It is standard practice to say that "lightspeed is the universe's *maximum speed limit.*" This is accurate if you are *only* thinking about the ***x, y, z*** spatial dimensions. But spacetime's speed of causality is spread through all of spacetime's ***x, y, z, t*** dimensions. And that full gallon of water representing spacetime's speed of causality is *always* in use. So if you look at all four dimensions, spacetime's speed of causality is not a *maximum speed limit*, spacetime's speed of causality is *spacetime's* ***only*** *rate of movement.* There is never more nor less of spacetime's speed of causality in play. It is always "all in." Every bit of matter or force in the universe is *always* moving at spacetime's speed of causality if you look at *all* of spacetime's ***x, y, z, t*** dimensions.

In this section, I've been careful to use the accurate, but clumsy, phrase "spacetime's speed of causality." That's all well and good, and helpful when digging into how this "speed" works. But it's a big, clunky phrase. So just remember, when I say "lightspeed," that's shorthand for "spacetime's speed of causality." "Lightspeed" is not simply a *speed* but more like the rate at which something that happens at one location in spacetime can affect something at a different location in spacetime — using all four ***x, y, z, t*** dimensions.

To sum up, lightspeed can be allocated *all* to a spatial dimension, as in the case of light itself. Or lightspeed can be allocated *all* to the time dimension, as when you are lounging on a park bench. Or lightspeed (spacetime's speed of causality) can be partly in the ***t*** jug and the rest in the ***x, y,*** and/or ***z*** dimensions — in the case of something that you are observing which is traveling by at close to lightspeed.

You Are Traveling At Lightspeed Through Time

How fast have you ever traveled? A commercial airliner can get up to about 0.25 miles per second. The SR-71 Blackbird reconnaissance jet could get up to 0.61 miles per second. As far as space travel goes, the ISS orbits at 4.75 miles per second. But light travels at 186,282 miles per second, so I consider it highly unlikely that you will ever personally encounter anything traveling in the ***x, y, z*** physical dimensions at close to lightspeed. Other than when you are observing light itself, of course. Or if you happen to work at a particle accelerator.

Every bit of reality is always moving at spacetime's speed of causality. But that doesn't mean that this "speed" is being used in the x, y, z physical dimensions. The bare fact is that you, as a person, spend only a tiny fraction of your speed of causality on moving through the x, y, z dimensions of physical space. Almost all of your speed of causality goes into t: the time dimension of spacetime. Follow this link if you want some more visualizations of this tradeoff in your speed of causality (*your* lightspeed) between movement through space and movement through time.[64] This presenter is a little quirky, but all of his videos are great! He summarizes the tradeoff between movement through spatial dimensions and movement through time as "Any gain in spatial speed is matched by a loss in temporal speed." This is what causes an object's time to slow down (dilate) when observing that object moving at close to the speed of causality in the physical x, y, z dimensions.

As noted above, a photon travels *at* spacetime's speed of causality through the physical x, y, z dimensions. None of a photon's speed of causality is left for the dimension t of time. A photon can travel across the entire universe and never experience time. Its internal clock is stopped.[65]

Full disclosure: quantum mechanical experiments show a couple of exceptions to spacetime's speed of causality, but that is at the mind-numbingly small scale of the quantum world. For the human scale of existence, we can ignore quantum effects. At our scale of existence, and from the size of dust motes to the size of galaxies and the entire universe, lightspeed rules. If you are curious about quantum mechanics, you can read about that in the appendexes.

Now here's the kicker: All of this is based on a frame of reference. Every frame of reference has its own reality. You, me, somebody on the International Space Station, an elementary particle whizzing through our solar system — every one of these has its own frame of reference and its own reality. How an object is moving *in your frame of reference* determines how you see that object's speed of causality spread between spacetime's x, y, z, t dimensions. From some other observer's frame of reference, it would be different, depending on how fast the observed object is moving *in that observer's frame of reference*. Every frame of reference is different, and they are all real.

That's Just Crazy Talk

Each unique prediction made by special relativity is, in itself, a hard pill to swallow. Taken all together, well, that's a lot of hard pills, and there'd better be some darned solid evidence to wash it down with or nobody would swallow all this stuff.

Is there more than the Michelson-Morley experiment to back up special relativity?

Why, yes! There is. I'll just mention a couple of examples.

Light Has Only One Speed

One prediction of special relativity that is hard to swallow is that the speed of light will be the same regardless of the speed of the object that created it (in your frame of reference).

Common sense tells us that if someone is standing in front of us and gently tosses us a ball, let's say at 5 mph, it will land gently in our hand. But if someone is driving by at 60 mph and just as gently lobs the ball to us, the ball will hit us pretty hard, at 65 mph. This common sense approach to how light *should* travel was called "the emission theory." Willem de Sitter was a Dutch scientist who specialized in physical cosmology. That's a fancy way of saying that he studied the large-scale structure of the universe. We are interested in him here because he proposed an experiment to test whether light travels as the emission theory states, or as special relativity states.

The De Sitter experiment,[66] proposed in 1913, looked at the light coming from double-star systems where *the stars rapidly rotate around each other in such a way that each star would at times be coming toward the Earth and at other times it would be moving away from the Earth.*

If special relativity were correct, light created by a quickly orbiting star would travel to Earth at a constant velocity regardless of whether the light was created while the star was moving toward the Earth or away from the Earth.

The competing, more common sense, emission theory said that the light created while the star was moving away from the Earth would travel to Earth slower than regular lightspeed, and the light created while the star was moving toward the Earth would travel to Earth faster than regular lightspeed. The fast light from one side of the orbit would reach Earth early, and the slow light from the other side of the orbit would reach Earth late. This would cause some well-defined optical illusions.

De Sitter's experiment tested special relativity's prediction that light *always* travels at the same speed regardless of any velocity difference between the source and the observer. De Sitter studied many double-star systems. The optical illusions predicted by the emission theory were not observed. Emission theory failed the test.

Light acts just as special relativity says it should. It always travels at our spacetime's speed of causality, aka lightspeed, regardless of the speed at which the object that created the light is traveling relative to the observer. Give Einstein a gold medal.

Length Contracts and Time Slows Down

Another couple of special relativity's hard pills to swallow are *length contraction* and *the slowing down of time (aka "time dilation")*. Length contraction states that when you observe an object speeding by at close to lightspeed, that object's length (along the direction of travel) will be compressed. When you observe an object speeding by at close to lightspeed, time will tick along very slowly for that object. That is to say, *in the observer's frame of reference*, the object's length is compressed and its passage of time slows down.

The **Rossi-Hall experiment** of 1941 was an early confirmation of length contraction and the slowing down of time.

This experiment observed a curious elementary particle called a muon. Muons are *very* unstable, and decay into other particles in a fraction of a second. When high-energy cosmic rays hit the top of the Earth's atmosphere, they bust apart whatever air molecule or atom they slam into. Among the typical pieces of shrapnel from such collisions are muons. These muons are propelled at close to lightspeed by these collisions, but even traveling that fast, they *should* decay into other particles before reaching the ground. At least according to Newtonian physics.

When instruments were built that could detect these elusive particles, the number of muons detected far exceeded the Newtonian expectations — but matched nicely with special relativity's predictions based on length contraction and the slowing of time for such fast-moving objects. OK. Give Einstein another gold medal. For a fuller description (and excellent illustrations) check out this link.[67]

The Kennedy-Thorndike experiment[68] of 1932 was basically a rerun of the Michelson-Morley experiment, with one important variation: the beam of light was split and recombined such that one path was shorter than the other. While the results of the Michelson-Morley experiment are explained by special relativity's *length contraction* alone, the results of the Kennedy-Thorndike experiment confirm both *length contraction* and *the slowing of time*. Chalk up one more in support of special relativity.

Other very nerdy experiments have also confirmed the slowing of time, but my favorite was the **Hafele-Keating experiment** of 1971.[69]

Atomic clocks are incredibly accurate. It is estimated that an atomic clock might get off by one second in a few tens of millions of years.

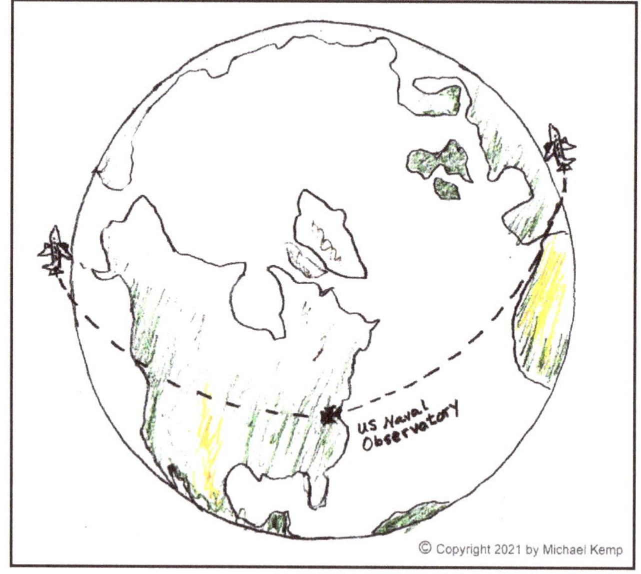

The Hafele-Keating experiment involved three sets of atomic clocks. One set stayed in the U.S. Naval Observatory, recording the flow of time at that spot on the rotating Earth. Another set of atomic clocks was flown around the Earth eastward. Since the Earth also rotates eastward, this resulted in a greater velocity relative to the clocks at the Observatory. Another set was flown around the Earth westward. Since this is in opposition to the rotation of the Earth, it resulted in a lesser velocity relative to the clocks at the Observatory. When the three sets of clocks were compared, the eastbound set had lost around 60 nanoseconds compared to the clocks that stayed in the Observatory, the westbound set had gained around 270 nanoseconds. This matches the changes in the flow of time predicted by the combination of special and general relativity. Chalk up yet another set of facts supporting relativity.

Suffice it to say, real-world observations have confirmed special relativity as a valid theory for how our universe actually functions.

The Warp Drive Loophole

OK, there is a loophole to all this "nothing can travel faster than light" stuff. The equations of *general* relativity indicate that, if you could protect your spaceship inside a small area of spacetime — a **warp bubble** — theoretically you could slide that bubble of spacetime along faster than light by compressing spacetime in front of the bubble and expanding spacetime behind the bubble. This is the fabled **warp drive**.

The catch? It appears that all of the proposed warp drive designs (at least the ones that would break the lightspeed barrier) would require an enormous amount of negative energy. And, if negative energy existed in any meaningful amount, it would *destroy the universe!* So, while warp drives are possible under Einstein's theory of general relativity, we haven't thought up a way to get there yet.

I know, I know, I'm just a party pooper. Follow this link for Sabine Hossenfelder's take on the state of warp drive research.[70]

Einstein Didn't Stop With ***Special*** Relativity

You may have noticed that special relativity is always framed in terms of a *constant* velocity between the observer and the subject. The essence of special relativity is that it compares two "inertial" (non-accelerating) frames of reference. It is this limitation to the special case of *non-accelerating frames of reference* that puts the "special" in special relativity. Einstein labored for a decade, presenting incremental insights, until one momentous November day in 1915 when his *general* theory of relativity was published. By incorporating acceleration into his understanding of the workings of the universe, Einstein overthrew and replaced our understanding of gravity. According to general relativity, gravity does not have to do with massive objects attracting each other. It has to do with massive objects bending the very fabric of the universe. And in general relativity, gravity and acceleration are one and the same.

I hope that special relativity has gotten you in a flexible frame of mind because in the next chapter we are going to get totally bent.

Review

"Relativity" means "how things look from a particular frame of reference." Every observer has their own unique frame of reference. A person standing on the shore, watching a ship smoothly making way on a calm sea has one frame of reference. A person standing in a closed cabin on that ship has another frame of reference. The idea of "standing still" means different things to these two people.

As long as there is a constant velocity between the observer and the subject, it's pretty straightforward to translate between their two frames of reference. Like in the illustration of the red sports car, the blue SUV, and the tree by the side of the road.

But the Michelson-Morley experiment and the Lorentz transformations showed that as you approach lightspeed, everything starts to get squishy. In the observer's frame of reference: (1) The subject's length shrinks in the subject's direction of travel (2) the subject's weight increases (3) The subject's passage of time slows down (dilates). In the subject's frame of reference, none of these effects occur. So your experience of reality is relative to your particular frame of reference. That's the "relative" in relativity. Like the direction of down, your reality depends on your location.

Surely this is just an illusion!

No. This has been physically tested. This is how our universe actually works.

And the word "lightspeed" is just shorthand for spacetime's "speed of causality." It is the fastest speed at which something that happens *over here* in spacetime can have an effect on something else *over there* in spacetime, when you take into account *all* of spacetime's four dimensions ***x, y, z, t***.

In Einstein's four-dimensional description of our universe, a specific time is just another type of location. Which makes intuitive sense. If I want to meet you for dinner, I'd better say where (which gives the three spatial dimensions of latitude, longitude, and height above sea level) plus a location in time (like 7:30 pm PST). Otherwise, we won't be at the same place at the same time. A specific time is a location in the time dimension.

In the thought experiment with four one-gallon jugs representing the four dimensions of spacetime, and one gallon of water representing spacetime's speed of causality, we saw how spacetime's speed of causality gets traded off between the spatial and time dimensions. In four-dimensional spacetime, everything travels at spacetime's speed of causality, you always use the full gallon of water between your four jugs. It's just that, if you are *not* moving close to lightspeed in the three spatial dimensions, then virtually all of your lightspeed gets used in moving you from the past through the present and into the future.

Given current technology, we never even get *close* to lightspeed in the spatial dimensions, so our speed of causality gets used in the time dimension, moving us from a location in the past, through our location in the present, and into a location in the future.

All this is incredibly nerdy. In our daily comings and goings, we really don't need to worry about all this stuff. The Newtonian Laws of Motion and Galilean relativity are close enough for everyday life.

But if you want to know how the real world *actually* works, "close enough" is weak tea. What you want is special relativity. It absolutely nails how space and time work — not just for planes, trains, and automobiles — but for the extremes. Like beams of light, muons created by cosmic rays, and particles in the Large Hadron Collider.

What has special relativity got to do with *why stuff falls down*? Special relativity challenges some of the assumptions made in Newtonian gravity. This pushed Einstein to formulate a new theory of gravity that takes into account all of the new experimental results, plus special relativity's elastic description of distance, time, and mass. In the next chapter we dig into this new description of gravity: general relativity.

Activity

Welcome to the Twins Paradox

Demonstrations of special relativity require equipment, precision, and huge amounts of energy that are simply not available in the home. So instead of a DIY activity for this chapter, I'll give you an extra-credit problem to explore.

With all the squeezing and stretching of space and time, it isn't too hard to come up with a thought experiment that makes special relativity look like it contradicts itself.

"The Twins Paradox" is a classic example.

I'll describe The Twins Paradox, let you wrack your brain over it, and then invite you to follow a couple of links that really dig into the meat of it and explain how the paradox isn't really a paradox.

Two twins, Bert and Ernie, live on Earth. Bert stays on Earth and Ernie hops a spaceship to Sirius (aka The Dog Star), which is 8.611 light-years away. The spaceship is traveling at 86.6% of the speed of light. When the spaceship gets to the star, it slingshots around the star and comes back to Earth, still going at the same speed.

Special relativity says that Bert, watching Ernie travel, will observe that Ernie's time is slowed down (dilated). To make the math easy, I'm setting the spaceship's speed at 86.6% of the speed of light because the Lorentz transformations give that as the speed where the flow of time is cut in half. While Bert watches Ernie travel out to Sirius and back, in Bert's frame of reference it takes Ernie about 10 years to get to Sirius and about 10 years to get back, for a total of about 20 years.

But Bert also observes Ernie's time slowing way down for Ernie. From Bert's frame of reference, Ernie will only age 5 years going out and 5 years coming back, for a total of 10 years. So Bert, who will have aged 20 years, expects Ernie to have only aged 10 years when he returns home. Bert expects to be 10 years older than Ernie when Ernie gets home.

Now here's the catch. Given that the heart of *relativity* is that everybody gets to have their own frame of reference, Ernie is perfectly justified in thinking that he (Ernie) is the stationary observer and Bert (and the Earth and Sirius) are the ones that are moving near lightspeed, first way off in one direction, then all the way back. Ernie is the stationary one and Bert moves back and forth (along with the Earth and Sirius). Since in Ernie's frame of reference, Bert (and the Earth) traveled away from him at 86.6% lightspeed and then returned at 86.6% lightspeed, it is Bert's clock that should run slow. It is Bert who should have only aged 10 years while Ernie aged the full 20 years.

So when they get back together, Bert thinks that he should be 10 years older than Ernie — and Ernie thinks that he should be 10 years older than Bert — because special relativity says that they both get to see things from their own frame of reference. From the observer's frame of reference (one of the twins), the other twin's time slows down when the velocity between them gets near lightspeed.

Obviously, it can't be both ways at once. They can't both become the older twin.

There's either some missing piece of the puzzle, some error in how we are looking at the situation, or some error in special relativity itself. Before you toss special relativity in the garbage bin, flip back to the *That's Just Crazy Talk* section of this chapter and re-read that sampling of physical experiments that have been done to confirm special relativity.

Give yourself a little time to think about this puzzle, and write down some thoughts about what might be going wrong here.

OK, I'll give you links to two explainers of this paradox. First, this excellent TED-Ed presentation by Amber Stuver: https://www.youtube.com/watch?v=h8GqaAp3cGs

And here's another good walkthrough by Henry Reich at the Minute Physics channel: https://www.youtube.com/watch?v=LKjaBPVtvms

Links

Biography
53. "Albert Einstein - Wikipedia." https://en.wikipedia.org/wiki/Albert_Einstein.

Deeper Dive
54. "Einstein's Theory of Special Relativity | Space." 30 Mar. 2017, https://www.space.com/36273-theory-special-relativity.html.

Biography
55. "Bernhard Riemann | | Britannica." https://www.britannica.com/biography/Bernhard-Riemann.

Biography

56. "Hermann Minkowski | German mathematician - Encyclopedia ...." 8 Jan. 2022, https://www.britannica.com/biography/Hermann-Minkowski.

Deeper Dive

57. "Annus Mirabilis papers - Wikipedia." https://en.wikipedia.org/wiki/Annus_Mirabilis_papers.

Deeper Dive

58. "Fission vs. Fusion – What's the Difference?" 30 Jan. 2013, https://nuclear.duke-energy.com/2013/01/30/fission-vs-fusion-whats-the-difference.

Great Video

59. "What Is Brownian Motion? | Properties of Matter YouTube." 30 May. 2013, https://www.youtube.com/watch?v=4m5JnJBq2AU.

60. "Galilean invariance - Wikipedia." https://en.wikipedia.org/wiki/Galilean_invariance.

Deeper Dive/Physics-Speak

61. "Inertial frame of reference - Wikipedia." https://en.wikipedia.org/wiki/Inertial_frame_of_reference.

62. "Galileo's ship - Wikipedia." https://en.wikipedia.org/wiki/Galileo%27s_ship.

Deeper Dive/Physics-Speak

63. "Relativity : the Special and General Theory by Albert Einstein." 7 Jan. 2022, http://www.gutenberg.org/ebooks/5001.

64. "We All Move At The Speed Of Light… Kind Of." https://www.youtube.com/watch?v=fB8eatgkOyM

Deeper Dive

65. "Does light experience time? - Phys.org." 8 May. 2014, https://phys.org/news/2014-05-does-light-experience-time.html.

66. "De Sitter double star experiment - Wikipedia." https://en.wikipedia.org/wiki/De_Sitter_double_star_experiment.

Great Video
67. "Impossible Muons - YouTube." 13 Nov. 2018, https://www.youtube.com/watch?v=rVzDP8SMhPo.

Deeper Dive/Physics-Speak
68. "Kennedy–Thorndike experiment - Wikipedia." https://en.wikipedia.org/wiki/Kennedy%E2%80%93Thorndike_experiment.

Deeper Dive
69. "Hafele–Keating experiment - Wikipedia." https://en.wikipedia.org/wiki/Hafele%E2%80%93Keating_experiment.

Great Video
70. "Are warp drives science now? - YouTube." 15 Jan. 2022, https://www.youtube.com/watch?v=YdVIBlyiyBA.

10. The Day the Universe Got Bent (General Relativity)

Student: "Dr. Einstein, Aren't these the same questions as last year's final exam?"
Einstein: "Yes; But this year the answers are different."

November 25th, 1915. That's the day that the universe got bent. Well OK, the universe did not change. It stayed what it always had been. But on that November day, Albert Einstein's definitive work on gravity was published: his general theory of relativity.

Now, don't get me wrong. Einstein wasn't the only one working along these lines. Most notably, a fellow physicist named David Hilbert, who communicated intensively with Einstein, developed the same theory of gravity in parallel with Einstein and presented his hypothesis at about the same time as Einstein. Check out this link[71] for a summary.

In this little book, I am crediting Einstein with the formulation of general relativity rather than Hilbert. Most sources agree, and Einstein has, after all, more gravitas.

We're finally getting to the heart of the matter. After two chapters of building up to it, now we get to Einstein's theory about *why stuff falls down*.

It's about time!

I mean that literally. In general relativity, the slowing of time creates gravity. Bent space is also essential in creating gravity. In fact, slowed time and bent space go hand in hand to get the job done. So it might be more appropriate to say that warped *spacetime* causes what we perceive as "gravity."

Are you ready? Let's get bent! According to general relativity, mass warps spacetime. How does this warped spacetime appear to us? Warped spacetime appears to us as an acceleration. That acceleration is what we call "gravity." That's general relativity, the short version.

To which I would expect you to say "Wait, what? Run that by me again, and explain it better this time!"

Let's set the stage. So where did *general* relativity come from? There were two things about special relativity that really tweaked Einstein's nose.

The first thing was that special relativity does not mesh with Newtonian gravity. In Newton's theory of gravity, changes in the gravitational attraction between massive objects occur instantaneously. For example, as one planet orbits the sun, the angle of its gravitational influence on another planet changes as they travel through their respective orbits. The strength of that gravitational attraction also changes as the distance between the two planets changes. According to Newtonian physics, the changes in the angle and strength of a planet's of gravity are felt instantaneously by another planet regardless of the distance between the two planets. In Newtonian physics, any changes in the gravitational influence of a massive object are felt by all other objects that have mass — to the farthest reaches of the universe. *Instantaneously*.

Einstein had gone "all in" with special relativity's prediction that nothing could travel faster than the speed of light (spacetime's speed of causality). Not even changes in the force of gravity. That is in direct conflict with Newton's theory of gravity.

The second thing was that special relativity is formulated for an observer and subject that are moving relative to each other at a *fixed, unchanging,* velocity. Another way to say this is that, in special relativity, both the observer and the subject are in *inertial* frames of reference. They are not accelerating. How would acceleration look in Einstein's elastic description of spacetime?

Speed, Velocity, Acceleration, and an Inertial Frame of Reference

Let's get some definitions straight before we go any further.

Speed is pretty simple. Speed is how far you travel in a given amount of time. The speed of light in a vacuum is 186,282 *miles* per *second*. How many *miles* per *hour* is that car going? 70 mph describes one particular speed. 25 mph describes a *different* speed. So far, so good.

Velocity is partly about speed and partly about direction. If a car is going 70 mph, what direction is it going? That makes a big difference! If you are trying to go from Los Angeles to San Francisco, you had better be headed Northwest, not East or South or any other direction. The velocity you should be traveling on the I-5 highway as you pass Bakersfield on the way from Los Angeles to San Francisco had better be 70 mph & direction Northwest (if you are traveling at the speed limit). You need both speed *and* direction to describe a velocity. 70 mph is not a velocity, it is a speed. Northwest is not a velocity, it is a direction. 70 mph & direction Northwest *is* a velocity. I hope that clears up any confusion about speed versus velocity.

Acceleration is a change in velocity. Remember that velocity is *both* speed and direction. Acceleration can be a change in speed and/or a change in direction. If a car changes speed, it is accelerating while it is changing speed. If a car changes direction, it is accelerating while it is changing direction.

For instance, there is an intersection of two highways near Pocatello Idaho that I have driven many times. Going East on Hwy 86, then taking the very gentle on-ramp to Hwy 15 heading North, you can keep a constant speed. But you are changing direction from East to North. That is an acceleration, even though your speed has not changed, your velocity has changed from 65 mph & direction East to 65 mph & direction North (if you are driving the speed limit).

A curious thing about acceleration is that *during* acceleration you can feel the pull of this change in velocity. If you are turning a sharp corner, you will feel a pull to the side away from your new direction. If you are speeding up, you will feel a pull backwards. If you are slowing down, you will feel a pull forward. This is due to the inertia of your body.

While you are at a constant velocity, your body is not pulled one way or the other. When your velocity changes, you feel the force of acceleration pulling against your inertia. You might recall Newton's first law of motion: **An object at rest will remain at rest unless acted upon by an external and unbalanced force. An object in motion will remain in motion unless acted upon by an external and unbalanced force.** An object's motion *is* its velocity, whether that velocity is 60 mph & direction North or 0 mph & direction None. An object "at rest" still has a velocity: 0 mph & direction None. It takes an "external and unbalanced force" to change an object's velocity. Inertia is both the tendency for an object at rest to stay at rest, *and* inertia is also the tendency for an object in motion to stay in motion (keeping the same speed and direction). If you are accelerated, either in speed or direction, you will experience the pull of that acceleration against your inertia. That's the pull you feel when rounding a sharp corner or braking suddenly. That pull is the force of acceleration working against your body's inertia.

An Inertial Frame of Reference. A "frame of reference" is the way the world looks (the coordinate system) of a particular observer. Remember that illustration in Chapter 9 with the blue SUV, the red sports car, and the tree by the side of the road? Each of those objects has its own frame of reference. The tree is not changing its velocity. The tree's frame of reference is "inertial" because it is not accelerating. As long as the blue SUV is traveling at a constant 60 mph without changing direction, the blue SUV's frame of reference is also "inertial." It is not accelerating. As long as the red sports car continues in the same direction at a constant 70 mph, then the red sports car's frame of reference is also "inertial." Neither the tree, the blue SUV, nor the red sports car will feel the pull of acceleration — because none of them are changing their velocity. They each have their own *inertial frame of reference*. Their inertial frames of reference are very different, but they are each real and valid.

If any one of them changed velocity: if the tree fell down; if the blue SUV swerved to the right; if the red sports car sped up to 80 mph — then that object would no longer have an inertial frame of reference while it was accelerating. It would feel the pull of the force of acceleration against its inertia, whether its acceleration was a change of speed or a change of direction, or both. If there is acceleration, then during that acceleration that object's frame of reference is no longer inertial.

Now that we've got those definitions under our belt, let's get back to the story.

Einstein wasn't content to rest on his laurels, as the old saying goes. He knew that something had to be done about the theory of gravity. He struggled with it until he had a thought that was both simple and strange.

You Don't Want to be in Einstein's Happiest Thought

In 1907, just two years after presenting his paper on special relativity, Einstein had his first inkling of a totally different way of thinking about gravity. He was musing about the personal frame of reference (the personal "system of coordinates") of someone who is falling off of a roof. He insisted that it was "the happiest thought of my life."

His "happiest thought" — his thought experiment — starts with a person standing on a roof. They feel the pressure of their weight as their feet are pressed against the roof. It is as if their body is being pulled downward against the roof. Once they fall off the roof, that pressure of their weight on their feet disappears. As they are falling, if they were to close their eyes, they would not feel pulled in any direction. They would feel that they were floating, weightless. They would not *feel* that they were accelerating. If anything, it would seem that it is the ground that is accelerating toward them.

An observer on the ground would have a different frame of reference. The observer on the ground would see the person falling, and accelerating, in the grip of gravity. This is our normal perspective on how things fall down. But Einstein chose to ponder the situation from the frame of reference of the person in freefall.

It is a very similar situation to the thought experiment where one person on a train drops a stone straight down and a person on the footpath sees it fall in a curve. Except that here we have — in the one mental image of an unfortunate person falling off of a roof — acceleration, gravity, and the hallmark of Einstein's relativity: two different observers with dueling frames of reference. One frame is accelerating relative to the other. But which one is "accelerating?"

It took Einstein eight more years to think through the implications of this "happiest thought," hammer out the details, and present his theory of general relativity. Einstein's theory of *why stuff falls down.*

Inertia, Acceleration, and Einstein's Gravity

According to Newton's ideas about inertia, Newton's famous apple should have stayed hanging in mid-air after it detached from its branch — unless some external force was being applied to it. In Newton's view of the world, that external force was the attraction between the mass of the apple and the mass of planet Earth. Newton saw gravity as an external force that acts on any object possessing mass. Newton saw the fall of the apple as the external force of gravity overcoming the apple's inertial tendency to stay right where it was.

Einstein turned all that on its head. Newton chose his personal frame of reference, sitting and watching the apple fall. Einstein chose the frame of reference of the person falling off the roof. The person falling off the roof feels weightless. They do not experience any external force. Einstein abandoned the idea of gravity being an external force, and rather, saw gravity as an acceleration in spacetime itself. The falling person is not being pushed or pulled, they are simply floating in the acceleration of a warped region of spacetime.

Einstein coined what he called **the equivalence principle**, which states that you can't tell the difference between a gravitational field and an acceleration that provides the same change in velocity. Gravity and acceleration are indistinguishable.

Elevators in Spaaaaaace!

Let's do a few thought experiments to highlight Einstein's hypothesis that gravity and acceleration are indistinguishable. You might think that this is just a curiosity, something that has no "real" consequence for how the fabric of spacetime actually works. If that's what you think, well, are you sitting down? These thought experiments end with a unique and testable prediction that shows just how "real" the equivalence principle is.

This set of thought experiments involves what happens inside of two elevator cabs. An elevator cab is the box that you step into to get lifted or lowered to your desired floor. In *all three* of these thought experiments, one elevator cab is just sitting still on the surface of the Earth. You could think of this as a closed elevator cab stuck on the ground floor of a building. There's a person inside who can observe anything that happens inside the cab. We'll call this the "Earth cab." The Earth cab is (obviously) subject to the gravitational field at the surface of the Earth.

The other elevator cab is way out in space, far away from any gravitational field. Let's just pretend that this cab is tightly sealed and has its own air supply. There's a person inside this cab too. They can observe anything that happens inside this cab. We'll call this the "space cab."

In both of these elevator cabs, there are no windows. The person inside each cab can only observe what happens inside their cab. *They have no knowledge of whether they are in the Earth cab or whether they are in the space cab.*

Now, let's review what Earth's surface gravity is like. If you drop something, that object will accelerate downwards, increasing its speed at a rate of 32 feet-per-second, every second. Ignoring air friction, if you drop something from a great height, at the end of the first second the dropped object will be going at a speed of 32 feet-per-second. At the end of the second second, it will be going 64 feet-per-second. At the end of the third second, it will be going 96 feet-per-second — and so on and so on, *increasing in speed by 32 feet-per-second for every second of free-fall.* This is the rate of acceleration that gravity creates near the surface of the Earth: 32 feet-per-second per second. This is often referred to as 32 feet per second squared, or 32 fps^2. This is what someone inside the Earth cab experiences if they drop something — anything they drop will accelerate at 32 feet per second squared until it hits the floor of the Earth cab.

Now, let's look at the space cab. In the three thought experiments that we are about to do, the space cab is *not* sitting still. It is imitating Earth's gravity by being pulled upward through space by a magical cable attached to the top of the space cab. For these three thought experiments, the space cab is being continuously accelerated at a rate equal to the acceleration of the Earth's gravity, increasing its speed by 32 feet-per-second every second. So at the end of the first second, the space cab is going at a speed of 32 feet-per-second. At the end of the second second, the space cab is going at 64 feet-per-second. At the end of the third second, the space cab is going at 96 feet-per-second. And so forth and so on…

Could passengers in these two cabs tell whether they were in the Earth cab (being subjected to Earth's gravity), or whether they were in the space cab (being pulled at a rate of acceleration which matches the force of the Earth's surface gravity)?

For the *first thought experiment* let's consider how weight is experienced inside the two elevator cabs. As noted in Chapter 5, the weight of an object depends on the mass of that object plus the gravitational field where the object is located.

We are all familiar with how weight works on the surface of the Earth, and by extension how it would work in the Earth cab. Standing inside the Earth cab, an observer will feel her weight holding her down to the floor of the cab. If she had a briefcase, and let go of it, it would drop to the floor of the Earth cab.

Now for the space cab. We have all experienced the effect when the automobile that we are riding in suddenly accelerates. We are pushed back against the car seat. The magic cable at the top of the space cab is accelerating the cab at 32 feet-per-second every second. The inertia of the occupant of the space cab, and any other objects that are inside the space cab, will cause them to be pulled in the opposite direction of the acceleration — toward the floor of the cab. The space cab's occupant would feel her weight holding her down to the floor of the cab. If she had a briefcase, and let go of it, it would drop to the floor of the space cab.

The experience of weight would be identical between the Earth cab and the space cab.

For the *second thought experiment*, let's harken back to a Galilean example of relativity. The one where a sailor in a closed cabin could toss a ball across the room, and would not be able to tell (from the movement of the ball) whether the ship he was on was at anchor, or smoothly making way on a calm sea. Update that for these elevator cabs, and the question becomes: If the occupants of each cab tossed a ball across the cab, could they tell by the movement of the ball, whether they were in the Earth cab or in the space cab?

Tossing a ball in the Earth cab is something we can all relate to. The ball will curve down as it crosses the cab, drawn down by the gravity of the Earth.

Tossing a ball in the space cab calls for a little deeper thought. From the point of view of an outside observer who is *not* accelerating, the ball will travel straight across the cab (the dashed green line in the next illustration) — *but* since the space cab is continuously accelerating as the ball crosses it, from the accelerating frame of reference *within the space cab*, the ball will curve down as it crosses the cab (the dashed red line in the next illustration).

As the next illustration shows, from the frame of reference of the observers in each cab, the ball curves just the same in the space cab as it does in the Earth cab. Earth cab on the left, space cab on the right.

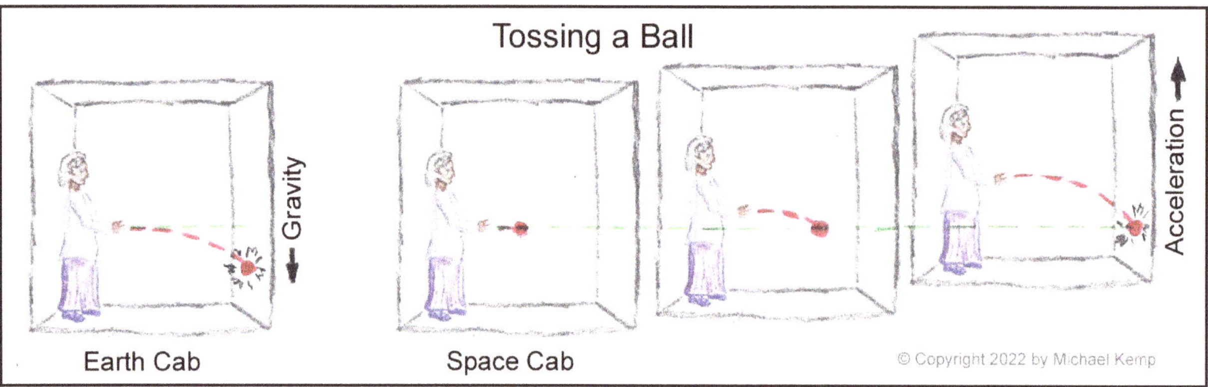

Once again, the observers inside both cabs will have identical experiences. They will be unable to tell whether they are in the Earth cab (experiencing what we would call gravity) or in the space cab (experiencing what we would call acceleration).

Here comes the moment of truth.

For the *third and final thought experiment*, the occupant of each cab shines a laser pointer across the cab, parallel to the floor.

In the continuously accelerating space cab, even though a beam of light travels enormously faster than tossing a ball, it still takes a fraction of a second to travel across the space cab (light travels at about one foot per nanosecond).

During that fraction of a second, the space cab will be continuously accelerating. Once again, the space cab will be going ever-so-slightly faster by the time the beam of light strikes the far wall of the cab.

As a consequence, the beam of light will hit the far wall of the cab just slightly lower than the height of the laser pointer where it started.

From the frame of reference of the observer in the space cab, the beam of light will bend down as it crosses the cab.

Einstein maintained, per the equivalence principle, that the same would be true of a beam of light crossing the Earth cab. Gravity would bend the beam of light down (just ever-so-slightly) as it traveled across the Earth cab.

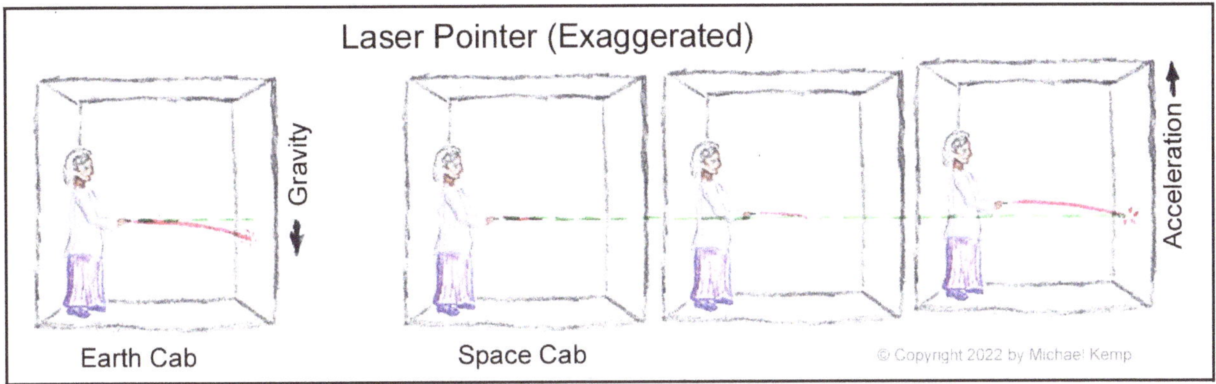

If Einstein's general relativity was right about gravity being warped spacetime, and *if* he was right about the warping of spacetime manifesting as an acceleration, then general relativity predicts that *gravity must bend light.*

Mic drop.

Nobody saw that coming!

Einstein's Orbit

Remember my illustration of propelling a baseball into a stable circular orbit in Chapter 5? In Newton's view of the world, the gravitational force pulling the baseball toward the Earth is balanced by the forward inertia of the baseball, to produce a circular orbit.

Next is the general relativity version of that illustration. Instead of a *force* of gravity, we have the *acceleration of warped spacetime.*

Instead of the forward inertia of the baseball, we have the *velocity* of the baseball. The velocity determines how much time the acceleration of warped spacetime has to act on our baseball. As before, at an elevation of about 250 miles (the orbit of the ISS), a velocity of 17,500 mph is just right to achieve a stable circular orbit.

For comparison, a passing photon travels at over 38,000 times the velocity of the ISS. The photon spends very little time in Earth's region of warped spacetime, and so its path is only very slightly bent by the acceleration of Earth's gravitational field.

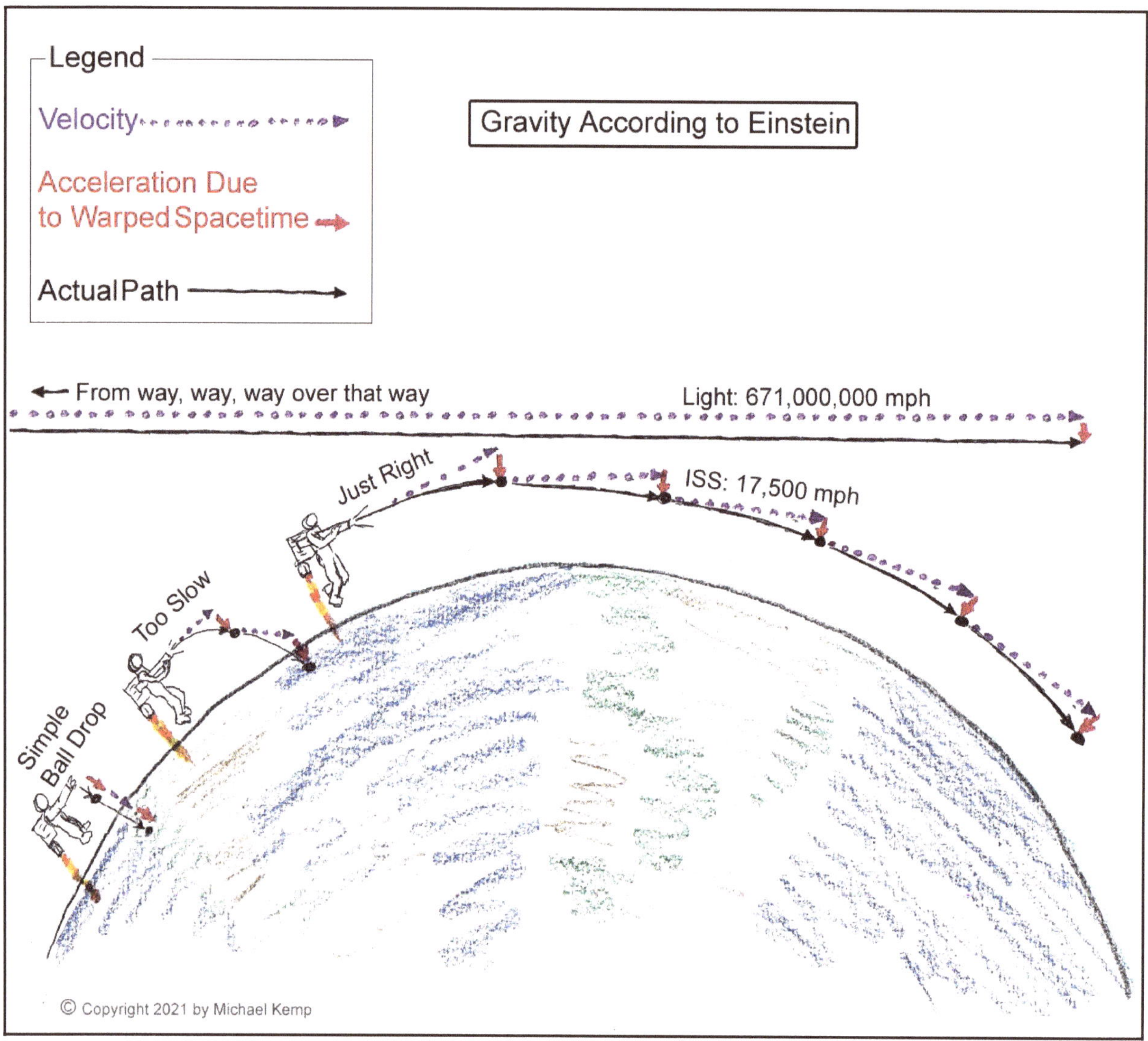

Mass Creates a Field of Warped Spacetime

According to general relativity, massive objects warp the spacetime around them. Actually, according to general relativity, any form of *energy* warps the spacetime around it, but mass equates to an incredible amount of energy. The famous formula **$E=mc^2$** roughly translates to "a tiny amount of mass has a ginormous amount of energy," it's just all bottled up. In fact, the energy bottled up in mass is *so ginormous* that for practical purposes we can ignore all the other types of energy that might warp space. At least as far as our day-to-day life is concerned.

This warping of spacetime manifests as an acceleration toward the massive object. This acceleration is in the fabric of spacetime itself. Anything that wanders into a warped region of spacetime will follow its warped ways and its path will bend toward the massive object (such as the Sun or the planet Earth). We call this warpage **gravity** or more accurately a **gravitational field**.

Einstein said that mass creates a gravitational field in much the same way that Michael Faraday described the magnetic field around a magnet. A magnet does not directly pull bits of iron to it. As Einstein puts it, the magnet *"calls into being something physically real in the space around it… what we call a 'magnetic field.'"* Bits of iron then respond to the magnetic field, not to the magnet itself. General relativity's gravity works in this same way.

A massive object warps the spacetime around it, creating an acceleration that is stronger near the massive object and weaker as one moves away from that object. In this way, a massive object does not directly pull other objects toward it. It creates a **gravitational field**. Anything that comes into this gravitational field follows spacetime's warped contours. This is why general relativity predicts that a massive object's gravitational field will even bend the path of light. Massless photons of light follow the bent contours of warped spacetime, just like anything else that comes near a massive object.

Remember the person falling from a roof in Einstein's "happiest thought?" From the perspective of general relativity, their fall was simply their personal system of coordinates finally syncing up with the warped spacetime around them. The warped spacetime of Earth's **gravitational field** always told their body to accelerate towards the Earth, but the roof had been in the way. Once the roof was no longer in the way, their natural state was to sync up with the warp of the local spacetime. That sounds all nice and peaceful, doesn't it? Still, it's gonna hurt when they hit the ground.

This talk about warped four-dimensional spacetime is pretty hard to visualize. What does a "warp" in the fabric of four-dimensional spacetime even look like? Our eyes don't see in four dimensions, so let's use an analogy that works in two and three dimensions.

Here's a thought experiment that uses a two-dimensional surface to explore the idea of warpage. A thin sheet of rubber is laying flat on a desk. It is not warped. There are two spots marked "A" and "B" on the sheet of rubber.

If an ant on that rubber sheet walks from A to B, the lowest-energy way is to walk in a straight line from A to B. On a flat two-dimensional surface, the shortest way from A to B is a straight line. It takes the least energy.

If that same rubber sheet is transformed into a hollow sphere, then it is still a two-dimensional surface, but it has been warped into a sphere. Sure, the sphere as a whole object is now three-dimensional, but the *surface* of the sphere is a warped two-dimensional surface. Now if the ant walks from A to B, the lowest-energy way for it to walk is in a curved line across the surface of the sphere. It's no longer a straight line, but it's still the lowest-energy path. So energy-efficient travel on a warped two-dimensional surface is not the straight lines that it would be on a flat two-dimensional sheet. Travel has gotten warped along with the two-dimensional sheet.

But spacetime is not two-dimensional. It has the three dimensions of space, plus the dimension of time. A "region" in this *spacetime* contains both spatial distance and time duration. It's quite a leap from a warped two-dimensional rubber sheet to warped four-dimensional spacetime. General relativity states that when massive objects warp spacetime, that warpage of spacetime is a region of spacetime that *is* an acceleration. It is an acceleration that carries along anything that wanders into it. This is what we experience as a gravitational field. This is what is holding you down to the Earth right now. You live in a gravitational field. You live in a region of spacetime that the mass of the Earth has warped.

This all sounds like crazy talk, but as you will see at the end of this chapter, this theory makes unique predictions that have been tested over and over. Our universe actually works this way.

If I hold a bowling ball chest high, I exert some effort to keep it stationary. That effort, that force, is stopping the bowling ball from following its lowest-energy path. That lowest-energy path, for the bowling ball, is to follow the Earth's warped spacetime. If I let go of the bowling ball, it follows that lowest-energy path. It accelerates downward at 32 feet-per-second per second. Until it hits the floor. At that point the floor is providing a force that keeps the bowling ball from going any further.

To return to our free-falling roofer: as they fall, they are following their lowest-energy path in warped spacetime from their former position on the roof to their upcoming encounter with the ground. In their free-fall, they are weightless. They feel no external force. They are floating in the accelerated spacetime of the Earth's gravitational field.

A totally different way to get an intuitive feel for how general relativity warps spacetime is to use a flexible sheet of fabric, stretched tightly over a frame, to represent the fabric of spacetime.

If you roll a marble across the fabric, it will follow a straight line. But then, if you place a bowling ball in the middle of the fabric, it will deform the fabric, creating a depression in the center. Now if you roll a marble across the fabric, with the bowling ball pulling down the center, the marble will no longer travel in a straight line — its path will bend toward the bowling ball. Its path has been accelerated in the direction of the bowling ball, because the mass of the bowling ball has deformed (warped) the fabric.

Follow this link[72] for a demonstration performed by the renowned physicist Brian Greene and his children. He's using a shot put rather than a bowling ball. They seem to be having fun!

Mass Slows Time — Just a Little

It's not just the spatial dimensions of spacetime that get warped by massive objects. Massive objects also slow down the time dimension of spacetime. This effect is greater near the massive object. The slowing of time near a massive object has a critical role to play in creating what we call gravity.

One analogy is to think of yourself walking at a steady pace on level ground. The mass of the Earth slows time, and it slows time *more* the closer to the Earth you are. Time flows more slowly at your feet than at your head. As you walk (in this exaggerated analogy) your head will get ahead of your feet. As your head gets farther ahead of your feet, your upper body swings forward from your feet and you fall down. Slower time at your feet than at your head has created gravity.

This analogy is pretty simplistic and slightly misleading, but it gives an approximation of how general relativity describes gravity. For a deeper dive, I recommend the video by Andrew Short, at this link.[73]

The amount that time gets slowed down in Earth's gravitational field is not noticeable by human standards. For instance, time runs slower at sea level than at the top of the mountains, but if you spent 10 years living on the beach, you would age a millionth of a second less than if you had spent those 10 years in a cabin, at 7,000 feet elevation up in the mountains. I don't think you'd notice.

The next illustration shows the slowing of time due to Earth's gravitational field.

- The clocks in this illustration are set at various elevations.
- The column of clocks on the left side of the illustration are all synchronized.
- The clocks on the right show how those same clocks' times have drifted apart in the course of 1,000 years.
- The clocks on the right show the number of seconds that time has run more slowly at each elevation relative to the deep space clock — at the moment that the deep space clock has reached 1,000 years.

Earth's Warping of Spacetime Slows Down Time
but It's Only a Few Seconds in 1,000 Years!

Set up clocks at different distances from the Earth.
Start the clocks running *all at the same time* (left side of this illustration) and then *stop ALL the clocks the moment when the Deep Space clock reaches 1,000 years* (right side of this illustration). The clocks that are closer to Earth will have fallen behind the Deep Space clock. The deeper in Earth's warped spacetime, the slower the clock.

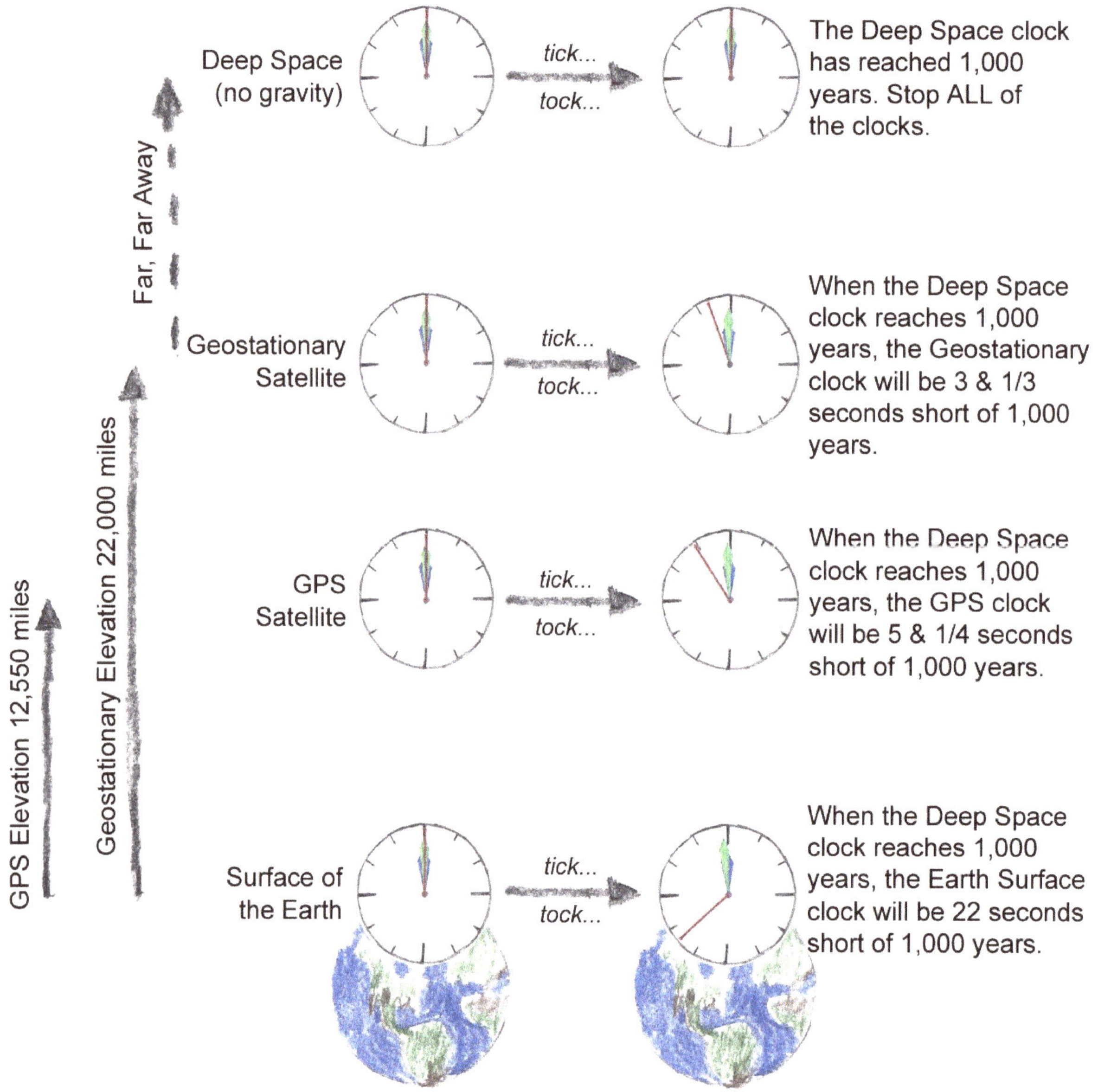

Deep Space (no gravity) — *tick... tock...* — The Deep Space clock has reached 1,000 years. Stop ALL of the clocks.

Far, Far Away

Geostationary Satellite — *tick... tock...* — When the Deep Space clock reaches 1,000 years, the Geostationary clock will be 3 & 1/3 seconds short of 1,000 years.

Geostationary Elevation 22,000 miles

GPS Satellite — *tick... tock...* — When the Deep Space clock reaches 1,000 years, the GPS clock will be 5 & 1/4 seconds short of 1,000 years.

GPS Elevation 12,550 miles

Surface of the Earth — *tick... tock...* — When the Deep Space clock reaches 1,000 years, the Earth Surface clock will be 22 seconds short of 1,000 years.

Note: The illustration above does not take into account time dilation due to special relativity. This is just the slowing of time due to general relativity. I've rounded the numbers to the nearest quarter or third of a second.

Let's get into the nitty-gritty. In the following sections, we will look at several unique predictions that general relativity makes, and how those predictions hold up to testing in the real world. Do the facts support this theory?

A Massless Difference

In Einstein's theory of general relativity, a massive object *bends spacetime itself,* warping it around the massive object. General relativity's prediction that a gravitational field will bend light is one of the places where Einstein's theory of gravity differs from Newton's theory. Light has no inertial mass. In Newton's theory of gravity, light would *not* be affected by gravity because light has no inertial mass.

According to general relativity, as light travels near a massive object, the path of the light will follow the warped contour of spacetime in the massive object's gravitational field. In Einstein's theory of gravity, the path that light takes *will* be affected by gravity. This is called **gravitational lensing** *(follow this link*[74] *for a couple of awesome Hubble Space Telescope photos of gravitational lensing).* The prediction of gravitational lensing gave scientists a way to test general relativity.

In the early 1900s, when Einstein proposed general relativity, they didn't have the Hubble Space Telescope to take photos of gravitational lensing, so they had to make do with a total eclipse of the Sun.

Here's how they tested gravitational lensing using a total eclipse of the Sun:

- Stars are fixed in their positions in our sky. As the Earth rotates, the stars *appear* to move across the heavens, but they are always in the same position relative to each other.

- As the Earth orbits the Sun, the Sun appears to move in front of different stars as the year progresses. The stars are in the same location. The Sun is in the same location. But as the Earth orbits around the Sun, different stars appear to move behind the Sun. This is called the ecliptic.[75]

- According to general relativity if the Earth is positioned in its orbit such that the Sun appears to be right next to a star (we'll call this the target star), light from the target star will be bent as it travels past the Sun. This will produce a slight optical illusion. The target star will appear displaced from its actual position in our sky, relative to other stars whose light does not travel near the Sun in order to reach the Earth.

- The actual position of this target star, relative to other stars, can be measured when the orbit of the Earth doesn't place the target star next to the Sun.

- Comparing how far the target star is from neighboring stars when it appears near the Sun as opposed to when it is not near the Sun will reveal whether or not its light gets bent when it appears near the Sun.

Here is a drawing of how the gravitational lensing of starlight produces the optical illusion of a star appearing to be in the wrong place in our sky.

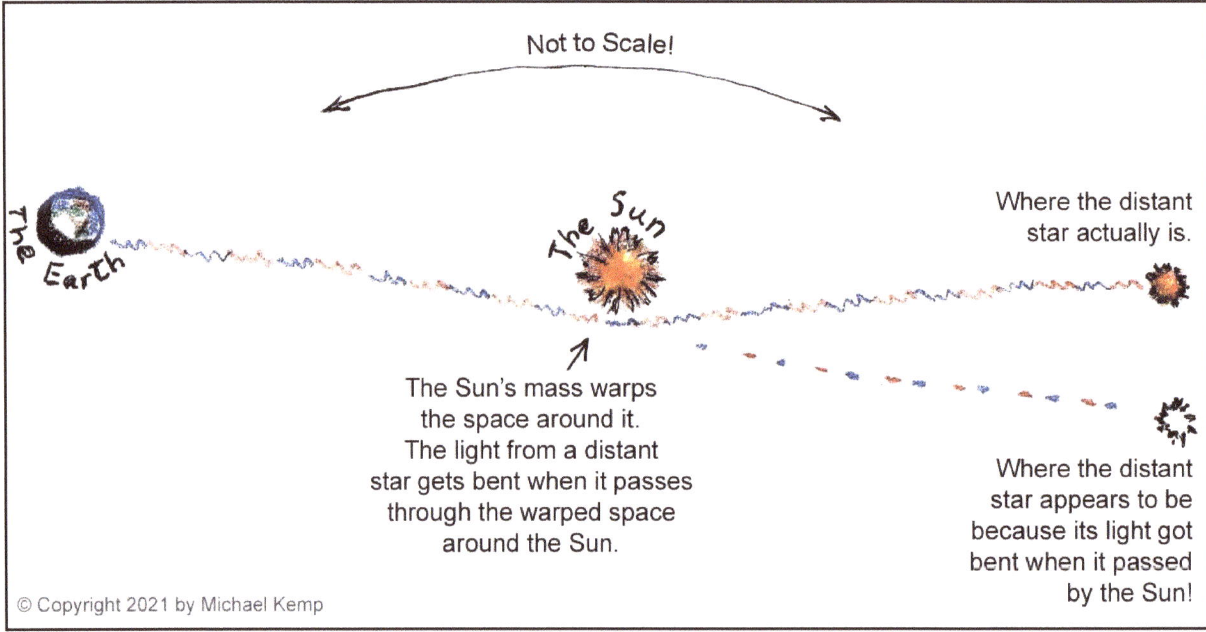

Of course, when the orbit of the Earth places our Sun next to the target star's position in our sky, the brightness of the Sun itself will overexpose any photograph of the target star, frustrating any attempt to test for this optical illusion. Except during a total eclipse of the Sun. During a total solar eclipse, a bright star next to the Sun can be photographed, and general relativity can be tested.

The Eddington Experiment provided just such a test in 1919.[76] Two expeditions were sent out to take photographs from two different points on the path of the total solar eclipse that occurred that year. Frank Dyson, the British Astronomer Royal, and Arthur Eddington, Director of the Cambridge Observatory, organized the expeditions.

One team went to Sobral, a town in northern Brazil. The other team, headed by Eddington himself, went to Principe, an island in the Gulf of Guinea off the West African coast. Both teams experienced severe challenges with equipment and with the weather, but in the end, they were both able to get a few usable photographs.

The photographs that the teams brought home showed that the star's light had indeed been bent by the Sun. This was obviously a big win for general relativity. No other theory predicted this result. Give Einstein yet another gold medal for that!

If you like British movies, you will enjoy the movie version of this expedition, which is titled *Einstein and Eddington*.[77] Heck, it's just fun to watch David Tennant (as Eddington), Andy Serkis (as Einstein), and the rest of the talented British actors in their supporting roles. And yes, the scriptwriters used some dramatic license, but hey, with that caveat, it's a fun watch!

In the years since the Eddington Experiment, improvements in telescopes have produced many examples of gravitational lensing. Here's a link to the EarthSky notes on gravitational lensing.[78]

Mercury's Motions

Another early test of general relativity was that it fixed a little problem with our calculations for the orbit of the planet Mercury.

Given not only the gravity of the Sun, but the gravity exerted on the planets by each other, the math to calculate the orbits of the planets gets pretty deep, but classical Newtonian physics was able to nail the orbits of the planets around the Sun. It was so accurate that when wobbles in the orbits of Neptune and Uranus were analyzed, Newtonian physics told astronomers where to look for the source of the gravitational anomaly. That is how Pluto was discovered.[79]

Did I say that Newtonian physics nailed the planets' orbits? Pretty much, except for Mercury.

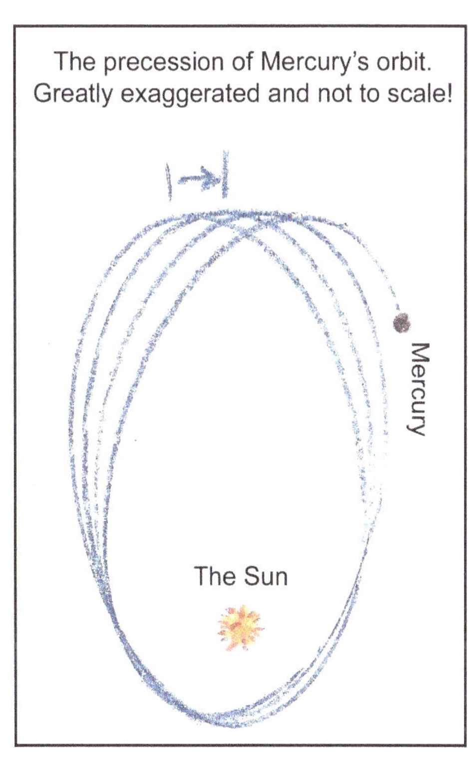

The precession of Mercury's orbit. Greatly exaggerated and not to scale!

Here's the deal. Planets do not orbit the Sun in perfect circles. The planets' orbits are oval-shaped. Due to the effect of each planet's gravity on the other planets, that oval gets altered a little bit each time the planet goes around the Sun. You could think of the oval itself rotating slowly as the planet repeats its orbits year after year. This rotation of the orbit is called precession.

In the 1800s astronomers realized that the precession of Mercury's orbit, using Newtonian physics, was just slightly off.

When calculations were re-run based on general relativity, they were found to be spot on.[80] So that was another persuasive piece of supporting evidence for general relativity.

Oh, good grief! Give Einstein another gold medal.

Go In And Never Come Out

The most dramatic prediction that general relativity made regarding the bending of light, is where you won't see any. Where you won't see any light, that is.

Theoretical physicists Karl Schwartzschild,[81] Hendrik Lorentz, and others glommed on to general relativity and, within a few months after its publication, were cranking out solutions to its equations. It took a constellation of physicists half a century to hammer it out, but the results were indisputable. General relativity predicts the objects that have become known as **black holes**.[82]

If mass can bend spacetime, and if that bending manifests as acceleration, and if that acceleration reaches the speed of light (spacetime's speed of causality), then such an incredibly massive object would create such a huge distortion in spacetime that nothing — not even light — could escape. It would appear as a black hole in our universe.

In honor of Karl Schwarzschild, the distance from the center of a black hole to the edge of the region of spacetime where nothing can escape, is known as the **Schwarzschild radius.**[83] This spherical surface, the boundary from which nothing can escape, is referred to as the **event horizon**.

At least nothing *directly* escapes from within the event horizon. See Hawking radiation[84] and the associated quantum entangled information leakage[85] if you want to dive into an area where quantum mechanics and general relativity meet amicably.

In 2006, an international collaboration called the Event Horizon Telescope was created to find direct evidence of these gravitational oubliettes. On April 10th, 2019, they successfully created an image of the matter orbiting at the edge of a black hole.[86]

This, and later observations, verify that the universe does indeed contain black holes. Gravity wells from which nothing can escape. Spacetime oubliettes, if you will.

As an aside, black holes were *not* a unique prediction of general relativity. Black holes had been predicted under Newtonian physics by John Michell way back in 1784, but that prediction was lost to the academic world. Michell's prediction of black holes was only rediscovered in the 1970s.

General relativity reignited interest in these gravitational monsters, and general relativity *did* add its own unique predictions as to what would be observed around the boundary of a black hole.[87]

General relativity also made the unique prediction that when two black holes spiral into each other, they would produce **gravitational waves** that would travel across the universe, rippling the fabric of spacetime. A century later, gravitational waves were detected. Follow this link[88] for a short video explanation.

Gravity Does the Wave

On September 14th, 2015, almost exactly a century after Einstein set out his theory of general relativity, the prediction that merging black holes would create gravitational waves across the fabric of spacetime was confirmed by the Laser Interferometer Gravitational-wave Observatory (LIGO).[89]

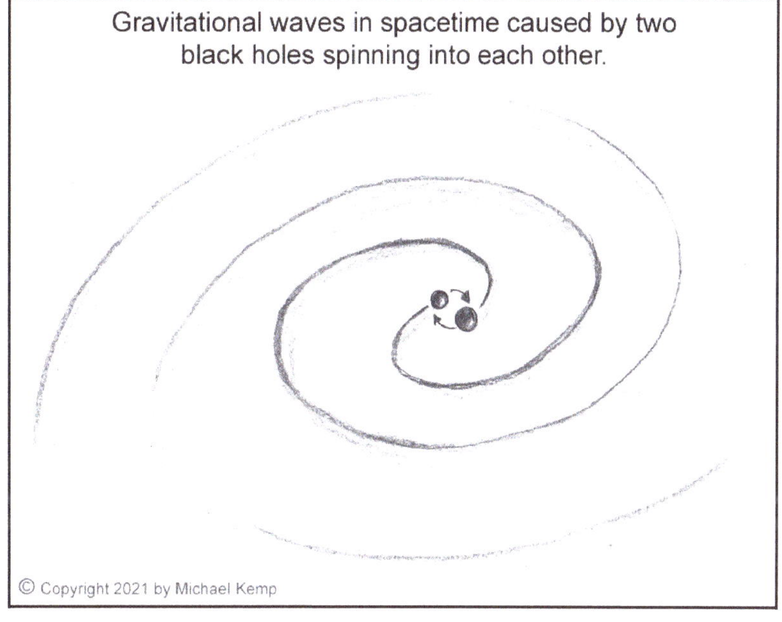

Gravitational waves in spacetime caused by two black holes spinning into each other.

General relativity predicts that in extreme gravitational situations, such as two black holes spiraling in to merge with each other, the enormous local warping of spacetime will send ripples across the universe.

From the data that LIGO detected, it is estimated that 1.3 billion years ago, two black holes with a combined mass of roughly 65 times that of our Sun, merged into a single, more massive black hole. This massive merger converted some of their mass into ripples in spacetime: gravitational waves. About three times the amount of the mass of our Sun was converted into the energy of these ripples in spacetime. Only general relativity predicted this. Once again, the universe behaves just as general relativity predicted that it would.

Einstein's gold medal count is getting ridiculous.

Many other gravitational wave events have been detected at the time of this writing.

A planned gravity wave detector (the Laser Interferometer Space Antenna, aka LISA) will use three satellites orbiting the Sun in a group. The satellites will keep a distance of 1.5 million miles from each other. They will stay in communication with each other via laser, detecting any distortions due to gravity waves by variations in the laser beams. This will dramatically increase the sensitivity available for detecting more gravitational wave events. But LISA won't be launched until the 2030s.

Your Head Runs Faster Than Your Feet

Can we test Einstein's assertion that a gravitational field will slow down time? Even the difference in elevation between a standing person's head and feet should be enough to make time run slower in their feet than in their head. It's not enough to make any meaningful difference in a human lifetime, but it should be measurable.

In 2010, the U.S. National Institute of Standards and Technology used atomic clocks to verify that the difference in the speed of time can be detected in as little as one foot of elevation.[90]

Yeah, yeah. Another gold medal for Einstein.

Want to know more? The unique predictions made by general relativity have been extensively tested. Follow this link for a list of experimental tests.[91]

General relativity has come through with flying colors.

Is That Your Final Answer?

Newtonian physics works really well. It meshes nicely with the common sense experience of our daily lives, and it does not hurt your brain. It saw nearly flawless service for over 200 years. However, as folks probed into the details of light, and made increasingly more accurate astronomical observations, Newtonian physics started looking less and less like the *Laws of Nature* and more and more like a very good approximation. Close, and good enough for normal everyday use, but not quite right.

In the 1800s and early 1900s, scientific experiments tore back the veil of a fixed, mechanical order that Newtonian physics had placed over our understanding of the universe. These experiments revealed the stranger nature of reality.

In dealing with *some* of these experimental results, special and general relativity were created. Special and general relativity describe how the universe behaves at the scale of human actions, planetary motions, and the universe as a whole. Special and general relativity accurately describe how light travels, how time gets slowed down, and *why stuff falls down*. Special and general relativity changed our classical idea of invariable length, universal time, unchanging mass, and Newtonian gravity into something much more flexible at high speeds, and bendy around massive objects. The equations of special and general relativity produce exquisitely reliable results.

So is general relativity our final answer to *why stuff falls down*?

Maybe, maybe not. The problem is that starting in the 1800s, we also looked a little too carefully at the smallest bits of our universe.

These *other* experimental results from the 1800s forced the formulation of new explanations of the smallest building blocks of our universe, a set of equations and theories that came to be known as **quantum mechanics**. Quantum mechanics revealed that at the atomic and subatomic scale, matter and force do not exist in continuously varying amounts, but are built up from tiny particles. Quantum particles of matter have specific amounts of mass and they interact with each other using specific amounts of force.

Experiment and theory also revealed that the position and momentum of these quantum particles are slightly uncertain. Uncertain, not because it is difficult to measure such small bits of matter and force (which it is), not because the act of measuring disturbs these tiny bits of existence (which it does), but because uncertainty is part of these particles' true nature.

When quantum mechanics became a viable theory, there immediately was a desire to meld the theories of the very large (special and general relativity) with the theory of the very small (quantum mechanics).

Special relativity was reconciled with quantum mechanics by the 1930s. But when attempts were made to extend the equations of *general* relativity down into the quantum realm, the results were a train wreck.

We know that general relativity is an accurate theory of how big stuff works. And it made a number of unique predictions which were later supported by experimental results. So general relativity is accurate and solid.

But quantum mechanics is accurate and solid too. Quantum mechanics has been supported by experimental results to a precision that is unmatched by any other theory in physics.

General relativity is excellent at describing gravity. Quantum mechanics has an unbeaten track record for predicting real-world experimental results for the smallest bits of matter and force. The fact that their equations do not fit together is a hint that we are missing something.

Remember, way back near the beginning of this little book, when I talked about rational thinking?

Rational thinking means that you compare one of your thoughts or beliefs to all of the other thoughts and beliefs that you hold to be true. If there are inconsistencies, you question the thoughts and beliefs that are in conflict.

There is no empirical evidence that either general relativity or quantum mechanics are flawed. Quite the opposite. And yet their formulas are in conflict with each other. It's a bit irrational to leave such a glaring conflict unresolved.

Given the clash between general relativity and quantum mechanics, physicists have been on a quest to find some description of nature that is at least as good at gravity as general relativity, and at least as good in the atomic and subatomic realm as quantum mechanics. In the next chapter, we will look at a couple of those attempts. It is a quest to find a Theory of Everything.

I hate to just breeze past quantum mechanics by saying "It makes folks wonder if general relativity is actually the final answer for gravity." I really want to give you an introduction to quantum mechanics, but even an "introduction" gets deep in a hurry. Quantum mechanics isn't about *why stuff falls* down, so I can't get away with sticking it in the middle of this book. But it's so important to modern science, and to the pursuit of a *new* theory of gravity, that I can't let it go. So just in case you are interested, I will add two appendixes at the end of this book, on the bizarre and fascinating world of quantum mechanics.

Review

Einstein's "happiest thought" turned Newton's idea of why stuff falls down on its head.

In Newton's theory of gravity, when an apple is parted from its branch, the apple's inertia (the tendency of an object that is sitting still to continue to sit still) is overcome by the force of attraction between the apple's mass and the mass of the Earth.

In Einstein's theory of gravity, once the apple is parted from the branch, the apple is freed from its bond to the branch and can follow the warp in spacetime created by the mass of the Earth. There is no force involved in falling.

One way to get an intuitive feel for Einstein's theory of gravity (general relativity), is to consider a gravitational field to be equivalent to an acceleration in spacetime. In fact,

Einstein called this "the equivalence principle." The section in this chapter titled *Elevators in Spaaaaaace!* was all about this equivalence principle.

General relativity builds on the elastic spacetime of special relativity. In special relativity, the relative velocity between the observer and the subject stretches and compresses time, space, and mass. In general relativity, mass and energy create a field of warped spacetime that is indistinguishable from acceleration. As the saying goes: matter tells spacetime how to curve, and curved spacetime tells matter how to move.

And it's not just space that gets bent, time gets slowed down too (not much, just a little).

Once again, this just sounds like crazy talk.

The problem is, it works.

One of the first observations to support general relativity was the bending of light from a distant star as it appeared near the Sun, during a total eclipse of the Sun. And then there was the Newtonian error in Mercury's orbit that was corrected by general relativity. Also, atomic clocks have measured the effect of gravity on time. And the gravitational waves predicted by general relativity have been observed. When folks have looked at how gravitational fields really function, general relativity absolutely nails it.

Activity

Don't Look at Me

Seriously, I got nothin'. Unless you happen to have a couple of atomic clocks that are reliable down to femtoseconds just laying around the house, or a telescope costing thousands of dollars and the patience to take dozens of hours of photographs of one particular gravitational lensing object and then stack those photographs up into a single consolidated image. Otherwise, I'm not thinking of any do-it-at-home activity that can demonstrate general relativity.

So the best I've got for you is to ask you to write down a few thoughts and make a few drawings or doodles about what we went over in this chapter. And review any of the links below that make you curious.

Links

Deeper Dive
71. "General relativity priority dispute - Wikipedia." https://en.wikipedia.org/wiki/General_relativity_priority_dispute.

Great Video
72. "Brian Greene Explores General Relativity in His Living Room." https://www.youtube.com/watch?v=uRijc-AN-F0.

Deeper Dive
73. How Time Dilation Causes Gravity, and How Inertia Works. https://www.youtube.com/watch?v=QNOFxmcECPQ.

Deeper Dive
74. "Gravitational Lensing." 30 May. 2019, https://hubblesite.org/contents/articles/gravitational-lensing.

Deeper Dive
75. "The ecliptic is the path of the sun - EarthSky." 27 Jan. 2017, https://earthsky.org/space/what-is-the-ecliptic/.

Deeper Dive
76. "Eddington experiment - Wikipedia." https://en.wikipedia.org/wiki/Eddington_experiment.

Great Video
77. "Einstein and Eddington (TV Movie 2008) - IMDb." https://www.imdb.com/title/tt0995036/.

Deeper Dive
78. "What is gravitational lensing? - EarthSky." 8 Sep. 2021, https://earthsky.org/space/what-is-gravitational-lensing-einstein-ring/.

Deeper Dive
79. "Pluto discovered – HISTORY." https://www.history.com/this-day-in-history/pluto-discovered.

Deeper Dive
80. "Precession of the perihelion of Mercury." https://aether.lbl.gov/www/classes/p10/gr/PrecessionperihelionMercury.htm.

Biography
81. "Karl Schwarzschild - Important Scientists" https://www.physicsoftheuniverse.com/scientists_schwarzschild.html.

Deeper Dive
82. "Black hole - Wikipedia."
https://en.wikipedia.org/wiki/Black_hole.

Deeper Dive
83. "Schwarzschild radius - Wikipedia."
https://en.wikipedia.org/wiki/Schwarzschild_radius.

Deeper Dive
84. "Hawking radiation - Wikipedia."
https://en.wikipedia.org/wiki/Hawking_radiation.

Deeper Dive
85. "Information May Leak from Black Holes at Dial-Up Speeds ...." 14 Mar. 2008,
https://www.scientificamerican.com/article/information-may-leak-from/.

Deeper Dive
86. "Wobbling Shadow of the M87* Black Hole | Event Horizon ...." 23 Sep. 2020,
https://eventhorizontelescope.org/blog/wobbling-shadow-m87-black-hole.

Deeper Dive
87. "Albert Einstein was right (again)" 9 Jul. 2021,
https://www.abc.net.au/news/science/2021-07-29/albert-einstein-astronomers-detect-light-behind-black-hole/100333436.

Great Video
88. "PBS Space Time | LIGO's First Detection of Gravitational Waves!." 24 Jul. 2016,
https://www.pbs.org/video/pbs-space-time-ligo/.

Deeper Dive
89. "Gravitational Waves Detected 100 Years After ... - LIGO Caltech."
11 Feb. 2016,
https://www.ligo.caltech.edu/news/ligo20160211.

Deeper Dive
90. "NIST Pair of Aluminum Atomic Clocks Reveal Einstein's ...." 23 Sep. 2010,
https://www.nist.gov/news-events/news/2010/09/nist-pair-aluminum-atomic-clocks-reveal-einsteins-relativity-personal-scale.

Deeper Dive
91. "Tests of general relativity - Wikipedia."
https://en.wikipedia.org/wiki/Tests_of_general_relativity.

11. The Quest for a Theory of Everything

Deep in the human unconscious is a pervasive need
for a logical universe that makes sense.
But the real universe is always
one step beyond logic.
~ Princess Irulan
(Dune, Frank Herbert)

General relativity and quantum mechanics were both the product of cold, hard facts. At the grand scale of humans, planets, and galaxies general relativity performs with stunning accuracy. At the scale of the unimaginably small, quantum mechanics also performs with stunning accuracy. Both theories made outlandish predictions that were later verified by experimental results. The problem is that their mathematical formulas do not play well together.

Attempts at combining general relativity with quantum mechanics' wave equations produce infinite results. And when an equation spits out infinite values it generally means that you are trying to apply the equation where it doesn't belong. General relativity and quantum mechanics are both descriptions of how the stuff of our universe behaves. The fact that they don't play nice together sure makes it seem like we're missing something.

Now, humans in general, and scientists in particular, like tidy answers. You can put it in as flowery and highfalutin and philosophical language as you please, but the simple fact is: we don't like it when the gears grind. We want all the parts of our understanding of the universe to mesh cleanly when they meet up. You could even make the argument that this is the essence of rational thought: your various theories about how the universe works should play nice together. When it became clear that general relativity and quantum mechanics did not play nice together, theoretical physicists spent decades trying to find some refinement or some mediating factor that would bridge the gap between the two theories. No such luck.

So physicists started looking for a new theory that would cover the entire range of existence from the unimaginably small to the mind-boggling immensity of the entire universe. Physicists started the search for *one theory* to explain all physical phenomena. They call this the search for a Theory Of Everything (TOE). Seriously, that's what they call their quest. It sounds like something Douglas Adams would have put in *The Hitchhiker's Guide to the Galaxy*. You just can't make this stuff up.

Some theorists approach this as a search for "quantum gravity." By adding gravity to quantum mechanics they would eliminate the need for general relativity. So if you see the phrase "quantum gravity," it's another way of saying "the quest for a Theory of Everything."

It should be noted that Nobel laureate Roger Penrose suggests that we should be looking for a general relativistic version of quantum mechanics, rather than a quantum version of gravity.

The quest for a Theory Of Everything is the search for a single theory to describe force, matter, space, and time — from the cosmic scale to the quantum realm. And by *force,* I mean the four forces of nature: electromagnetism, gravity, and the so-called "strong" and "weak" forces. The strong and weak forces only operate at atomic distances. Some versions even include dark matter and dark energy. On the other hand, they aren't talking about life, love, thought, or consciousness. So they're not really talking about a theory of *everything* everything. Just a theory of everything that we, in this day and age, consider to be *physical* phenomena.

There are many TOE candidates, but currently, the two most prominent candidates for the TOE are **string theory** and **loop quantum gravity**. These are both referred to as theories, but according to the generally accepted scientific definition of "theory," you don't get to call your hypothesis a theory until at least one of its unique predictions has been verified by experiment.[92] While both of these TOE hypotheses do a decent job of "predicting" experimental results that are already explained by quantum mechanics and general relativity, neither of them has made *unique* new predictions that our current level of technology has allowed us to test. So I think they are hypotheses, not theories. But scientists and science publications call them "theories" anyway. Go figure.

What follows is a brief description of these two TOE candidates. Many physicists feel that one or the other of these candidates are failed dead ends. Many other physicists insist that they ain't dead yet, give 'em more time! I present them here to give you a glimpse into the *type of thinking* that has gone into solving the riddle of unifying general relativity and quantum mechanics.

And yes, other hypotheses have been proposed to reconcile gravity with quantum mechanics (Google **modified gravity**, **causal dynamical triangulation,** or **entropic emergent gravity** if you want to see some of these other proposals), but **string theory** and **loop quantum gravity** are two of the most popular proposals for describing gravity in a quantum mechanical universe.

So far in this little book, everything has either been historical or supported by facts (aka observed phenomena, or experimental results). On the other hand, *the ideas in this chapter are speculative*. Take them with a grain of salt.

TOE #1: A Song of Superstrings

What if the ancient belief that the universe was filled with "the music of the spheres" wasn't so far-fetched after all? What if vibration was *the* primal source of existence?

Religious and philosophical traditions from the Bhagavad Gita to the Kabbalists associate sound (vibration) with the creation of existence. Or to put it in Christian terms, what if you take "the Word" to be the vibration that commands all things into existence:

In the beginning was the Word, and the Word was with God, and the Word was God.
The same was in the beginning with God.
All things were made by him;
and without him was not any thing made that was made. ~ John 1:1-3

So, it's not surprising that particle physicists decided to take the idea out for a spin. Quantum mechanics reveals that matter and force are composed of individual particles. Quantum mechanics only describes these particles as having *attributes*. Prime examples of these attributes are mass, electrical charge, and the so-called "spin" and "color" attributes. Quantum mechanics does not describe the *structure* of quantum particles.

In the early interpretations of quantum mechanics, quantum particles were treated as infinitely small zero-dimensional points. In this view, they have no structure. *But, what if the structure of a quantum particle is like a vibrating string?*[93]

What if each quantum particle was an unimaginably small rubber-band-like energy string vibrating in a tiny bundle? What if the mode of its vibration determined the string's mass, spin, and charge? By having a particular mass, spin, and charge the vibrating string *becomes* one of the quantum particles that we have discovered and named. What if the string's possible modes of vibration were constrained in such a way that it could *only* produce the quantum particles that we see in our universe?

Physicists put forward a variety of hypotheses about the nature of quantum particles from the 1920s onward. In the late 1960s, there was the first hint of particles as tiny vibrating strings. The early version of string theory was able to explain only force particles (such as photons), and did not include matter particles (such as quarks and electrons). In the 1970s and '80s, a concept known as supersymmetry[94] gained acceptance among some particle physicists. Supersymmetry expanded the number of elementary particles described by quantum mechanics by proposing that the matter and force particles of quantum mechanics have much more massive "symmetrical" twins.

String theory physicists glommed onto supersymmetry, and thanks to supersymmetry, string theory was reborn as "superstring theory," although it's still usually referred to as just "string theory." This change not only pulled matter (quarks, electrons, etc.) into the string theory fold, but it also addressed other riddles such as why particles have their particular inertial masses.

Also in the 1970s John H. Schwarz and Joël Scherk (and separately Tamiaki Yoneya) made another big splash: they used the equations of quantum mechanics to show that string theory predicts the graviton, an elementary particle that would carry the force of gravity. In this scenario, gravitons would be exchanged to create the gravitational field. That may sound strange, but according to quantum mechanics, all the other forces are mediated by particles. In the bizarre world of quantum mechanics, having the basic forces of nature be mediated by particles is the norm. So the string theory description of why stuff falls down is: the elementary particle called the graviton does it!

String Theory Needs Extra Dimensions

We are accustomed to thinking of the physical world as existing in three spatial dimensions and time (x, y, z, t). By the early 1990s, string theory had incorporated the concept of multiple *hidden* spatial dimensions. These extra spatial dimensions were needed mathematically, in order for the strings to have the proper resonating chamber in which to create the particles that we observe.

Since we only experience three physical dimensions and time, then according to string theory these *extra* physical dimensions must be hidden from us. The proposal is that each of these extra dimensions is bent. In fact, each of these extra dimensions is bent so tightly that it is tied up in knots that are so small that they are completely undetectable.

With the addition of multiple hidden dimensions, string theory could now predict all of the known elementary particles, plus that particle to transmit the force of gravity, the graviton.

Admittedly, standard quantum mechanics also predicted all of the known elementary particles, but string theory's addition of the graviton is a nice touch. So far, we have not been able to detect the graviton in any experimental device.

The incorporation of supersymmetry and hidden dimensions into string theory was called "The First Superstring Revolution."

Initially, it was thought that six hidden dimensions would produce the correct results. Adding that to the three dimensions of space and the dimension of time brings the grand total to 10 dimensions. Needless to say, the math to calculate the vibrating string's "notes" which were possible in this number of dimensions, including quantum effects, is daunting. In fact, the equations are so complex that they are beyond the current capability of humans and computers combined. Estimates are used instead, mathematical approximations, so to speak.

In approximating the mathematics of 10-dimensional string theory, string theory split up into five separate versions. Then one day a flurry of online postings and emails resulted in "The Second Superstring Revolution." If you added one more hidden dimension (incorporating what is called the 11D supergravity hypothesis), all five versions of string theory (plus 11D supergravity) could be treated as different manifestations of one central consolidated theory, which they called M-theory.

There is an old parable about five blind men and an elephant. When the group of blind men encounters the elephant, each man comes in contact with a separate part of the elephant. One, holding a tusk, declares that the elephant is a massive spear. Another, touching a leg, declares that the elephant is a tree trunk. Yet another, encountering the side of the elephant, declares that it is a wall. The fourth, holding the trunk, insists that it is a snake. The fifth, holding an ear, declares that the elephant is a large fan. They are each correct, but they all lack the full picture. Like the parable of the blind men and the elephant, the five separate descriptions of string theory turn out to be different appendages of a more inclusive beast, a "theory" that incorporates one more hidden dimension: M-theory.

Depending on who you ask, the "M" in M-theory means "Mystery", or "Mother," or "Magic," or "Membrane." I like Membrane, because string theory describes these hidden dimensions as being knotted up into specific shapes, the best visualizations of which look like twisted-up bits of membrane. For some reason, when physicists talk about membranes in more than two dimensions, they call them **branes** (not to be confused with brains).

These twisted-up branes of hidden dimensions can be represented by what is called Calabi-Yau manifolds.[95] It is impossible to render a seven-dimensional membrane on the two-dimensional page of a book or on a two-dimensional digital screen, but the image in this link will give you some idea:

https://aether.lbl.gov/bccp/Images/calabi-grid.gif

The energy and vibration of a string are channeled by the particular Calabi-Yau shape that it happens to be tangled around and that energy and vibration produces a particular quantum particle: photon, electron, quark, neutrino, whatever. The vibration and the shape together make the particle. These are the notes in the music of creation, according to string theory. The vibrating strings call into existence all of the elementary particles of matter and force.

String theory enthusiasts point out that it not only "predicts" already proven elementary particles, but it also offers insights into black holes, gravity, and the big bang.

But string theory does have issues.

For one thing, the math is astronomically difficult. As far as I can determine, nobody really understands the math for making clear-cut string calculations. Approximations are used instead. For another, supersymmetry, which saved string theory, predicts the existence of "supersymmetric particles" that are paired up with regular quantum particles of matter and force. Given the predicted mass of these supersymmetric particles, the Large Hadron Collider really should have found some of them by now. But it hasn't.

So string theory still seems to be struggling, half a century after vibrating strings of existence first became a twinkle in a physicist's eye.

On the other hand, string theory hasn't been specifically *dis*proved. There are quite a number of respected physicists who continue to explore and have confidence in string theory.

TOE #2: Loopy Gravity

Loop quantum gravity[96] (LQG) looks at reality differently.

I should note that there are several different interpretations of LQG. The interpretation I'll mention here sees space, time, and gravity as constructs, as artifacts produced by the deeper nature of existence.

What if reality is simply quantum particles and the events and relationships that they engage in? What if space, time, and gravity as we experience them, emerge from all those myriad quantum-scale events and relationships?

Consider what you see when looking at a high-resolution digital picture. A high-resolution picture emerges from all of its individual pixels. The whole picture is created from how each pixel is colored and each pixel's relationship to the pixels around it. What if space, time, and gravity are emergent properties of the events and relationships of all the uncountable quantum particles in the universe? Just like a high-resolution picture emerges from all the properties and relationships of its individual pixels.

Carlo Rovelli,[97] a prominent LQG theoretical physicist, writes: "In the elementary grammar of the world, there is neither space nor time — only processes that transform physical quantities from one to another, from which it is possible to calculate probabilities and relations." And again: "In a theory of this kind, time and space are no longer containers or general forms of the world. They are approximations of a quantum dynamic that in itself knows neither space nor time. There are only events and relations. It is the world without time of elementary physics." In other words, the quantum bits of our universe interact with each other, and at the human scale, those interactions produce the *experience* of space, time, and gravity.

LQG describes our sense of time as an artifact of our macro-level experience, and of the progression of **entropy**.[98]

Entropy can be described as a measure of disorder. As it applies to our perception of time, entropy can be seen as a one-way agent of change. According to the second law of thermodynamics, the total entropy of a closed system can never be reduced. Disorder in a closed system can only stay static or increase.

For example, let's consider a coffee cup and the coffee in it to be the closed system that we are studying. If we start with a freshly poured cup of hot coffee in a cold cup, entropy is low because our closed system has some order to it: hot coffee, and cold cup. As the coffee warms the cup it also cools down the coffee until the coffee and cup are at the same temperature. Our closed system of cup-and-coffee now has a higher level of entropy because there is less difference (less order) between the parts of the system.

Burning a piece of firewood is another example of increasing entropy. Before the fire, the wood is a complex structure of bark, cambium, and heartwood composed of individual cells with internal structures. As the fire burns the wood, all this structure (all this order) is destroyed. In the end, if the fire is efficient, you are left with the heat and light produced, plus homogenous gasses and ash. So at the end of the fire, you have a much lower state of order. You have increased entropy.

Suffice it to say that entropy is a one-way process. LQG proposes that entropy gives time its arrow from the past to the future. And in fact, the arrow of time is created by the process of entropy.

At the quantum scale, this version of LQG also considers gravity as an interaction between individual quantum particles. The complex webs of interactions between quantum particles are called spin networks.[99] A spin network is a snapshot of a single state at the quantum level. As interactions develop, the resulting, ever-changing web of interactions is called "spin foam."

Clear as mud, right?

OK. All of this spin network/spin foam stuff is ridiculously nerdy and abstract. The point is that this is the level at which LQG describes gravity. Right down in the quantum grains of the universe. Gravity begins in the interactions and relationships between all those quantum particles.

As for why stuff falls down, it would take a phenomenal number of quantum particles within a spin network/spin foam to produce anything that we would notice as "gravity" at the human scale of existence. But then again there are many, many elementary particles of matter and of force that are required to form and maintain a single atom, and a phenomenal number of atoms required to make a gravitational body that we would notice — such as the Earth.

So in LQG, gravity is an artifact of the relationships and interactions of the quantum particles that reality is built from. Our experience of time and space are just emergent properties, constructs: artifacts of our blurred perception of the relationships and interactions of an unfathomable number of quantum particles, and the process of entropy.

This may be a hazy description, but it's as good as I can give you. If you want to dig deeper, I suggest Carlo Ravelli's books. Or whatever has come out lately on the subject.

Review

So there you have a couple of different hypotheses that strive to bridge quantum mechanical observations with the observations supporting general relativity. There are many other hypotheses being floated around to do this job, but as far as I can tell none of them pass muster to be truly called a scientific theory. So I'll just use air quotes around String "Theory."

Once again, if you want my introduction to quantum mechanics, please refer to the appendexes.

Well now. That wraps up our journey through the centuries of refinement of our understanding of *why stuff falls down*.

In the next (brief) chapter we will sum up where this journey has gotten us. Then I can't help but leave a very short postscript with a brain-teaser or two.

Links

Great Video

92. "Fact vs. Theory vs. Hypothesis vs. Law… EXPLAINED!"
https://www.youtube.com/watch?v=lqk3TKuGNBA

Deeper Dive
93. "String theory - Wikipedia." https://en.wikipedia.org/wiki/String_theory.

Physics-Speak
94. "Supersymmetry | CERN." https://home.cern/science/physics/supersymmetry.

95. "Calabi–Yau manifold - Wikipedia." https://en.wikipedia.org/wiki/Calabi%E2%80%93Yau_manifold.

Physics-Speak
96. "Loop quantum gravity - Wikipedia." https://en.wikipedia.org/wiki/Loop_quantum_gravity.

97. "Carlo Rovelli - Wikipedia." https://en.wikipedia.org/wiki/Carlo_Rovelli.

Physics-Speak
98. "Entropy - Wikipedia." https://en.wikipedia.org/wiki/Entropy.

99. "The fabric of space: spin networks « Einstein-Online." https://www.einstein-online.info/en/spotlight/spin_networks/.

12. Are We There Yet?

"The Universe is under no obligation to make sense to you."
~ Neil deGrasse Tyson

So why *does* stuff fall down?

If I drop a feather and a hammer, I know which one will hit the ground first. Aristotle was right.

Unless I'm dropping a couple of lead balls, then air resistance doesn't slow the lighter ball down enough to matter. Same thing if I drop stuff in a vacuum, then (as you've seen if you've followed this link[100] or this link[101]) they hit the ground at the same time. So Galileo and Newton were right.

Unless you look at the gravitational lensing of starlight[102] or need to explain the precession of the planet Mercury's orbit.[103] Then Einstein's general relativity is right.

Unless you take the experimental evidence that our universe is constructed of infinitesimally tiny quantum particles — and feel the need for a single unifying Theory Of Everything. Then maybe (per string "theory") gravity is transmitted by gravitons which are a natural function of tiny vibrating strings that make up the universe **or** maybe (per loop quantum gravity) stuff falls down as a result of the interrelationships between the unimaginable number of incomprehensibly small quantum particles that make up our universe.

So are we there yet?

No. Not really.

General relativity gives a *why* for gravity that exquisitely nails *how* gravity truly behaves in our universe. But there are nagging questions at the edges of our vision that remain unanswered.

In the end, will the human mind prove capable of comprehending the incredible structure of our universe?

That seems to be an open question.

Maybe we just haven't looked at it from the right angle yet.

Until then, we can do quite nicely in our daily lives by using Newton's theory of gravity, even knowing that it is just a very close approximation.

When you have to be really truly on target, from how an apple falls, to how light bends near massive objects, to how merging black holes create gravity waves across spacetime, Einstein's general relativity does the job right.

Just don't try using general relativity down in the quantum realm.

I can live with that.

Can you?

Links

Great Video
100. "Apollo 15: The Hammer-Feather drop - YouTube." 5 Jan. 2015, https://www.youtube.com/watch?v=oYEgdZ3iEKA.

Great Video
101. "Brian Cox visits the world's biggest vacuum YouTube." 24 Oct. 2014 https://www.youtube.com/watch?v=E43-CfukEgs.

Deeper Dive
102. "Gravitational lens - Wikipedia." https://en.wikipedia.org/wiki/Gravitational_lens.

Deeper Dive
103. "Precession of the perihelion of Mercury." https://aether.lbl.gov/www/classes/p10/gr/PrecessionperihelionMercury.htm.

P.S. The Pull of the Dark Side

"Not only is the Universe stranger than we think,
it is stranger than we can think."
~ Werner Heisenberg

From the revelations of Isaac Newton to those of Albert Einstein, we have recognized the role that ordinary matter plays in creating gravity. By "ordinary matter" I mean the stuff that we can touch and see, that we can drink, and breathe — and even the atoms and elementary particles that we can detect through the use of particle accelerators. I mean all the matter that we know about through *direct* observation.

So massive objects make all the gravity, right?

Well, I'd hate to leave you with a false impression…

The 2nd Biggest Known Unknown in the Universe

This little book is about gravity. As such, I just *have* to include the most powerful and pervasive source of gravity that we have found — and our ignorance about what in the universe is causing it.

It turns out that ordinary matter only accounts for 1/6th of the observed gravity in our universe.

How do you "observe" gravity?

For one thing, you can measure how fast stars orbit around the center of their galaxy. The more gravity the galaxy is generating, the faster its stars have to be going to keep from getting sucked closer to the galactic core. So as telescopes got more powerful and stars were observed to be orbiting their galactic core faster than they had any reason to be, the question was "Where's all this extra gravity coming from?"

Another way to observe gravity is to measure how entire galaxies move relative to each other. A Bulgarian-born astronomer named Fritz Zwicky[104] spotted this type of anomaly. Zwicky even coined a term for the unseen source of this extra gravity, he called it **dark matter**. Given the immature state of astronomy in the early 1900s, his numbers were off, but later studies in the 1970s and 80s (by American astronomers Vera Rubin and Kent Ford) supported his observation and gave more reliable numbers for the extra gravity. Zwicky's general conclusion was correct: the galaxies he studied had a lot more gravity than could be accounted for by ordinary matter.

Early on, it was thought that the extra gravity would eventually be accounted for by ordinary matter that was not picked up by the telescopes of their times: planets, dust, comets, stars that had gone dark, and so forth. The problem was that, as the technology of astronomy got better, the problem of dark matter got worse. It turned out that there was no way that all of the planets, stars that had gone dark, dust, comets, and so forth, (even throwing in the discoveries about black holes[105]) could account for the colossal amount of extra gravity that the swiftly orbiting stars and galaxies were trapped in.

And remember the gravitational lensing that causes the light from distant stars and galaxies to be bent as it goes past a massive object? You can calculate how much gravity that massive object has by measuring how much the light gets bent. As ever-better deep space telescopes produced ever sharper images of gravitational lensing, the calculations only confirmed and refined the earlier observations. Something really heavy is happening out there in the dark.

According to the calculations, the cause of 5/6ths of the observable gravity remains invisible to our instruments.

Another curiosity is that this dark matter doesn't clump together like ordinary matter. We can tell from the distribution of the gravity that dark matter creates, that it is diffuse. I'm talking *really* diffuse. The dark matter of a galaxy is spread out in clouds that are much larger than the ordinary matter of that galaxy.

But is it really "matter?" I like Neal deGrasse Tyson's suggestion. He proposes calling whatever it is "dark gravity," because that's all we really know about it. Whatever it is, it reveals itself to us as extra gravity. For a much better description than I can give, watch Dr. Tyson lay it out in the video on this link.[106] If you don't like the term dark gravity, he suggests, we could just call it Fred.

If dark matter *is* matter, physicists have suggested that it could be produced by so-far undetected *extremely massive* elementary particles. They've even come up with a whimsical name for them: Weakly Interacting Massive Particles (WIMPs).

Other astrophysicists propose the exact opposite: a so-far-undetected extremely *low-mass* particle that they've nicknamed an "axion." Let's spend a moment considering these proposed, extremely low-mass, axion particles.

The quantum wave nature of matter does not manifest at the human scale of existence. The quantum wave nature of matter becomes more dominant as you look at smaller and smaller masses.

If the proposed axion particles have a mass so small that it would take one hundred trillion of them to have as much mass as one electron, well then, the axion's quantum wave would be gigantic. The quantum wave of a single axion particle would be the size of a galaxy. Uncountable numbers of these axion particles would *merge* in a galaxy-sized mass. Axions would become a nigh-undetectable superfluid around galaxies.

Simulations of these ultra-light axion particles match our observations of dark matter at least as well as simulations of the ultra-heavy WIMPs.

But both WIMPs and axions are only hypothesized. Neither have been detected.

Maybe the extra gravity doesn't come from matter at all: maybe it comes from something we haven't even conceived of. Or maybe we could explain away "dark matter" by adding some sort of modification to our theories about the gravity that is produced by normal matter? It is only at galactic distances that we have observed the phenomena that we attribute to dark matter. Maybe regular old gravity just operates differently at those huge distances. Some folks who make hypotheses about what they call modified gravity, or Modified Newtonian Dynamics (MoND for short) claim that with the proper MoND formulas, there is no need to hypothesize any dark matter.

Long story short: since this little book is about gravity, I'm honor-bound to tell you that we don't have a clue where 5/6ths of the gravity in our universe comes from. But we know from observing stars, galaxies, and the bending of light that it exists. A known unknown, so to speak.

Just so you don't throw up your hands and walk away, keep in mind that when an apple parts from its branch or a planet orbits the Sun, ordinary matter and general relativity totally account for how the apple falls and how the planet orbits. So what we *do* know about gravity is good. But it's not complete.

It just goes to show you that there is a lot more to be discovered out there.

Oh, and why did I call dark matter "the ***2nd*** biggest known unknown?" There is another force at work in our universe that has been labeled **dark energy**. Dark energy is the name given to whatever it is that is causing the otherwise unexplained (and increasing) rate at which galaxies are flying apart from each other. Judging by this increasing rate of universal expansion, dark energy deserves to be called ***the*** biggest known unknown in the universe.

But this little book is about stuff falling down, not about whether the universe will eventually be ripped apart by dark energy.[107] So I just thought I'd mention it in passing.

And that brings us to the end of our story. Well, not to the end so much as to "as far as we've gotten so far." From "stuff falls down to rejoin its proper sphere" through "mass attracts mass" to "mass tells spacetime how to bend and bent spacetime tells mass how to move."

As we have gone along, each theory has described the facts, the observed phenomena, better than the theories that came before it. The stunning leap of the scientific revolution has taken us from the Middle Ages to lasers, rockets, the end of smallpox, and the creation of nuclear weapons. All in 500 years. Now 500 years may seem like a long time, but given the age of our earliest written records (at least 3,500 years ago), given the age of the beginning of agriculture (almost 12,000 years ago), given the age of the earliest bones of our ancestors (hundreds of thousands of years ago)… 500 years is ***nothing***.

And I might add that we, with the curiosity to explore and dream, and with the scientific method to weed out defective ideas, will most certainly continue to expand our knowledge — and who knows — maybe even our wisdom.

We are a curious breed!

Michael Kemp

Links

Biography
104. "Fritz Zwicky | Swiss scientist | Britannica." https://www.britannica.com/biography/Fritz-Zwicky.

Great Video
105. "What If Dark Matter Is Just Black Holes? - YouTube." 13 Apr. 2021, https://www.youtube.com/watch?v=qy8MdewY_TY.

Great Video
106. "Why dark matter is a misnomer, with Neil deGrasse Tyson - Big Think." 6 Jun. 2017, https://bigthink.com/videos/neil-degrasse-tyson-dark-matter-is-a-misnomer.

Great Video
107. "What Is The Big Rip? - Universe Today." 18 Dec. 2013, https://www.youtube.com/watch?v=QRb3EiOkxrY.

Appendix A:

Breaking the Universe to Bits (Quantum Mechanics Part 1)

"Those who are not shocked when they first come across quantum theory cannot possibly have understood it." ~ Niels Bohr

Empirical evidence, experimental results, aka observed phenomena — cold, hard facts — all tell us that Einstein's general relativity is the last word on *why stuff falls down*. So what's left to say?

And why in the world am I bothering you with **quantum mechanics** in a book that's supposed to be about gravity?

I'm bothering you with quantum mechanics because there is a clash between quantum mechanics and general relativity. This makes us wonder whether general relativity might ***not*** be the final answer to *why stuff falls down*.

When physicists tried to make the equations of general relativity (GR) mesh with the equations of quantum mechanics (QM), the equations turned around and slapped the physicists in the face. Today, after a century of attempts to reconcile GR with QM, it's still a train wreck.

Maybe we could dismiss these inconsistencies if the predictions of QM were sloppy, questionable, or inaccurate. We could say "Oh, don't worry about GR not meshing with QM. QM is so speculative, don't give it another thought."

But that's hardly the case. We've built massive particle accelerators just to see if we could support QM, or disprove QM. Quantum mechanics has come through with flying colors.

So yes, I'm going to bother you with quantum mechanics. I've devoted two appendixes to the strange, bizarre, and wonder-filled world of QM.

In Appendixes A and B, we are entering the realm of the smallest bits of matter and force, from which everything that we experience in the universe is built. You, me, the Earth, the Sun, stars, galaxies, ice cream, lip gloss. All are built from the bits of matter and force, the **subatomic particles** and **elementary particles** described by QM.

Why did I break the story of quantum mechanics into two separate appendixes, two separate stories? Because one story (Appendix A) is about how we learned that matter is, indeed, constructed from atoms, subatomic particles, and elementary particles, and that even the forces of nature come in *distinct elementary particles*. The other story (Appendix B) is about how we learned just *how absurdly these elementary particles behave*.

In the 1800s our understanding of how things work at the smallest scale of existence fell apart. Experimental results supported the ancient Greek theory of atoms, but they didn't stop there.

One line of exploration that was taken in the 1800s had to do with the relationship between heat and light. That is what Appendix A is all about.

Now let's rewind back to the year 1800 and follow the trail of breadcrumbs that led to **quantum mechanics**. While based on experiment and observation, quantum mechanics defies common sense. The results of the experiments that we are about to explore broke our understanding of the structure of the universe to bits. Unimaginably tiny bits. Quantum bits.

At our human level of experience, the brightness of light can be continuously varied, brighter or dimmer. This matches the Newtonian view of the forces of nature. The results of studies in the 1800s that dug into the relationship between light and heat produced results that violated the Newtonian idea that light is continuously variable. This became known as **The Ultraviolet Catastrophe**. Let's follow the steps that led up to The Ultraviolet Catastrophe, and how this forced us to recognize that light is made up of really really tiny bits. These tiny bits of light, these elementary particles of light, are called **photons**.

The Heat of Light

While Isaac Newton was onboard with the atomic description of matter, Newtonian physics assumes that space, time, and the forces of nature can vary in a continuous manner. That is to say, in Newtonian physics there is no piece of space, time, or the forces of nature that is so small that it cannot be cut down even smaller.

Take distance, for example. You can measure an inch and divide it in half to create 1/2", then you can divide that to make 1/4", then 1/8", then 1/16", 1/32", 1/64", and so on. It sure looks like you could continue cutting distance in half for as long as your tools for dividing and measuring allow you to continue. The same goes for heat, light, and the forces of nature, as well as for time itself — in the Newtonian worldview space, time, and the forces of nature are continuous, unbroken, and endless. You can always cut a smaller piece.

Experiments performed in the 1800s destroyed these presumptions.

So what broke the rules? Well, for one thing, people looked a little too closely at the way in which heat can generate light, and light can generate heat.

So how can the study of light and heat have blasted our concept of the universe to bits? A man named Hershel set the stage in 1800 with his studies of how light creates heat. It took a century, and a man named John Strutt (aka Baron Rayleigh), to stroll out onto this prepared stage with a catastrophic equation that blew up Newton's conception of continuously variable forces of nature. Then it took a man named Planck to sweep up those broken pieces of the old, Newtonian physics, and replace them with a new, granular, description of our universe. So let's start with Herschel.

Sir William Herschel[108] got interested in the properties of light through his devotion to astronomy. He mapped the stars in the sky, ground his own telescope lenses, discovered the planet Uranus, and put forward hypotheses about nebulae and the evolution of stars. For his work, he was appointed as the king's astronomer, awarded a yearly pension, and knighted.

Herschel's investigation of light's spectrum and the heat transmitted by different colors of light is what is relevant to us here. He had noticed that when viewing the Sun through differently colored filters, he felt different levels of heat in the light that came through each filter. This got him curious about how much heat was generated by the various colors of the rainbow.

Herschel used a prism to break out the rainbow of colors in sunlight, then he recorded the effect of each color on thermometers whose bulbs had been painted black. To his surprise, an elevated temperature was also recorded beyond the red end of the spectrum, where no light was visible. This study of color and heat was published in 1800.[109]

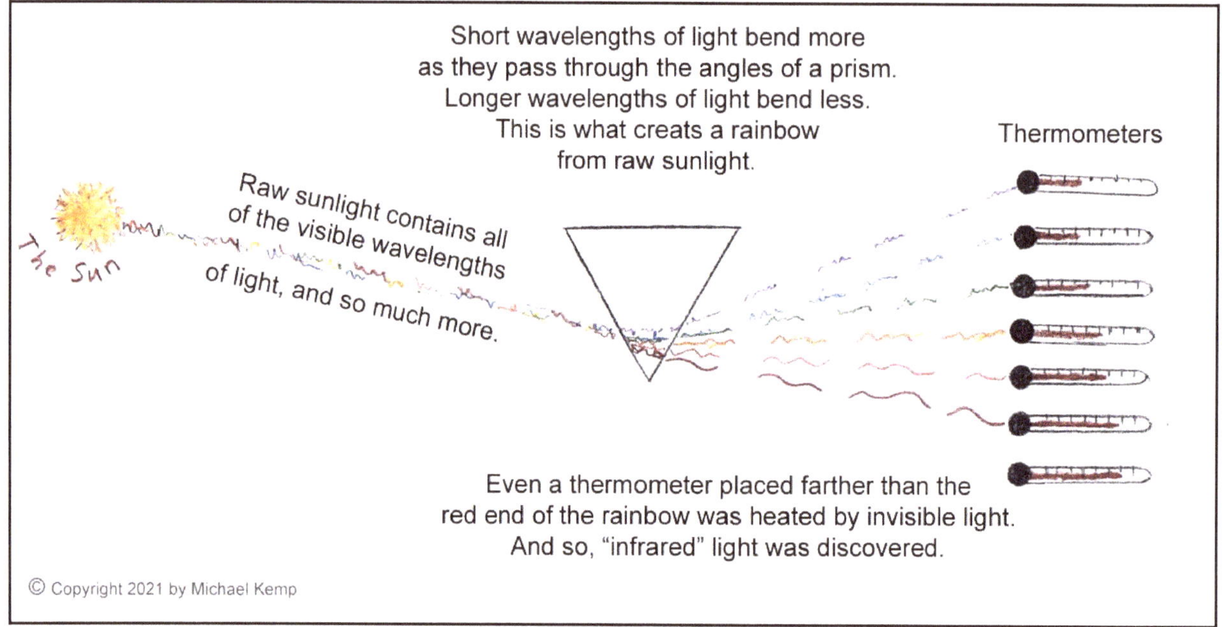

This was a big clue that visible light was only a portion of a larger phenomenon, a phenomenon that we now call the electromagnetic spectrum.

Some years later it was discovered that light comes in a range of "wavelengths." You might remember this from *Secrets Hiding in the Light* in Chapter 8, and a guy named James Clerk Maxwell and his electromagnetic spectrum.

The invisible heat that Herschel found was dubbed "radiant heat" and it was recognized to be the same as the heat one feels coming directly from a wood stove, a bonfire, or from the Sun itself.

This is ***not*** the hot air curling above the wood stove, but rays of heat radiating directly from the heat source. When you feel your face warmed by the heat of a wood stove, a bonfire, or the Sun itself, you might notice that you can block this heat by putting your hand in front of your face. Your hand is blocking rays of heat that radiate directly from the heat source. Most of that radiant heat is from the invisible light that Herschel discovered when he placed a thermometer beyond the red end of the rainbow.

These days we also call radiant heat "infrared" radiation because it is "farther than" red in the rainbow created by a prism. Infrared light has a wavelength that is longer than human eyes can catch and process. But snakes, mosquitoes, and frogs can see some of the infrared portion of the spectrum.

It was not long before it was shown that invisible light also exists beyond the opposite end of the rainbow of colors, beyond violet. And hence the name ultraviolet. Ultraviolet light has a wavelength that is shorter than human eyes can catch and process. But birds and most mammals including dogs and cats, as well as butterflies and bees can see some of the ultraviolet portion of the spectrum.

When studies were made to find how much energy the Sun puts out across its full spectrum of light, it was found that sunlight is most intense across the portion of the spectrum that makes up visible light. The peak energy was found to be at a wavelength of about 0.5 microns. This wavelength is a blue-green color. Visible light is comprised of wavelengths from around 0.7 microns at the red end of the rainbow to around 0.38 microns at the violet end of the rainbow.

But the Sun doesn't stop with visible light. Some of the Sun's energy reaches us in the ultraviolet portion of the spectrum. An even greater amount reaches us in the infrared portion of the spectrum. See the graph below to get an idea of the shape of this lopsided bell curve.

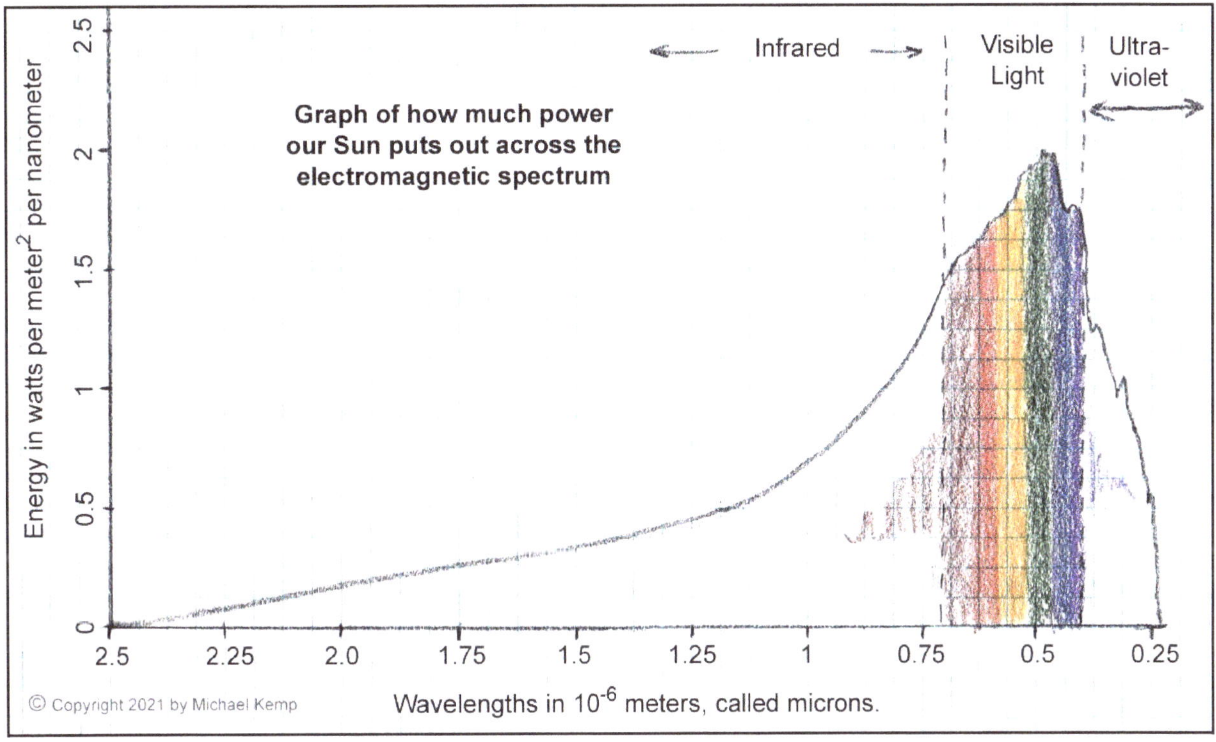

Granular Light

As more experiments were conducted on the spectrum of light, it became obvious that if you want to study light in a controlled way, you need a light source that you can vary at will. Something that would be brighter than a candle, and not as dependent on the weather and the time of day as the Sun.

As the 1800s rolled on, incandescent light became the go-to way to generate light in the laboratory. Incandescent light is created by heating an object until it produces light.

It became popular to take something black, like a piece of iron, gradually heat it, and observe what color it turned at which heat. You may have seen this yourself if you have ever left an iron poker in a fire long enough. The end of the poker will go from black to red to orange.

So, incandescent light is generated by heating an object. Visible light, as we saw in Chapter 8, is only a small fraction of the electromagnetic spectrum. Some of the electromagnetic radiation being created by heating objects is visible light, but a lot of this incandescent "light" is outside of what the human eye can catch and process.

The term **blackbody radiation**[110] was coined in the mid-1800s by the German physicist Gustav Kirchhoff[111] to describe this electromagnetic radiation which is created by heat. It is called "blackbody" because scientists wanted to start with a black object — one that was not giving off any light of any color — and see how much energy would be produced, and at what wavelengths, as this "black" object was heated to higher and higher temperatures.

Blackbody radiation could be studied more precisely if the heat used, and the light being created could be carefully controlled and isolated. The experimental setup for this became known as a **blackbody device**. The idealized version of a blackbody device is to have a hollow box that is totally black on the inside. The box has a hole through which visible light, and electromagnetic radiation that is not visible to the human eye, could escape. The observer can then record the wavelengths and intensities of the light generated inside the box. External heat is applied to the blackbody device until the interior of the box is so hot that it produces photons of visible and non-visible light. And yes, the box had better be made out of metal or firebrick or some such material that can survive high temperatures without catching on fire or melting down.

One of the many things that Gustav Kirchhoff proposed was that a blackbody device, being heated, will reach a **thermal equilibrium** by dumping out the same amount of energy as is being pumped into it.

A blackbody device can dump energy out by heating the air around it. It can also dump energy out by creating photons. With a constant input of heat energy, the blackbody device will heat up and dump energy out more rapidly as it gets hotter. When the blackbody device gets hot enough to dump energy back out equal to the speed and amount of energy being pumped into it, it achieves thermal equilibrium and will remain at a constant temperature for as long as the input of energy remains constant.

But, if the rate at which energy is being pumped into the blackbody device heats it beyond what it can physically handle, it will keep getting hotter until it melts down into a lump of slag. There's always that.

If the blackbody device is very well insulated, it cannot dump energy out by heating air and it will achieve thermal equilibrium by dumping all of the incoming energy back out by creating photons. This is where a blackbody device gets useful for comparing the amount of energy going into it versus the intensities and wavelengths of the photons coming out of it.

The photons being created by a heated blackbody device do not come at a single wavelength but are spread over a range of wavelengths in the electromagnetic spectrum. If you graph the intensity of the light (vertically) against its wavelength (horizontally) you will get a lopsided bell shape, similar to the illustration we just saw of the Sun's energy output at various wavelengths. The cooler the blackbody device, the more this range of wavelengths drifts to the long-wavelength, low-frequency, end of the spectrum. That puts it toward the red end of the spectrum and beyond, into the infrared. The hotter the blackbody device, the shorter the wavelengths, and the higher the frequencies of the electromagnetic radiation that it dumps out. That pushes its wavelengths toward the violet end of the spectrum and beyond, into the ultraviolet.

All waves have both frequency and wavelength. If you stand at one spot watching a water wave go by you, a wave's frequency is how many wave peaks pass by you in a second. On the other hand, its wavelength is the distance between one peak and the next. The higher the frequency the shorter the wavelength, and the lower the frequency the longer the wavelength.

Just for fun, I'll note that the energy/wavelength graph of the photons created by a human being peaks around 9.5 microns, solidly in the infrared area. Sadly, this means that we cannot see ourselves glow in the dark. Unless we use night vision goggles to translate infrared light into a color that we *can* see.

Mosquitoes, with their differently constructed eyes, unfortunately, *can* see us glow in the dark. Rotten luck.

Welcome to 1900, when Baron Rayleigh[112] (aka John William Strutt), with an assist from Sir James Jeans,[113] developed a formula from Newtonian physics to describe how much energy a blackbody device should produce at each wavelength. This formula shook Newtonian physics to its roots. But first, a bit about the good Baron.

If you've ever wondered *why* a clear sky is blue, Baron Rayleigh is the one who figured it out. He discovered that light is scattered (deflected from its path) when it encounters particles that are smaller than its own wavelength. And the shorter the wavelength of the light, the more it is scattered. Red light is scattered least, and blue and violet are scattered most.

When sunlight strikes Earth's atmosphere, the majority of the light barrels straight through the air to reach the Earth's surface without being scattered by little particles. That's why the Sun itself appears so bright. However, some sunlight *is* scattered by the atmosphere.

Since blue light is scattered most, the blue portion of sunlight will spray out in all directions from whatever point it hits the Earth's atmosphere. This is why the entire sky in every direction appears blue on a sunny day. Blue light from the Sun striking anywhere in the atmosphere will be scattered in all directions, including toward where you happen to be standing. Hence, blue sky.

Red light is scattered the least and has to travel through a lot of the Earth's atmosphere before it's going to spread out just a little. At high noon, sunlight only has to travel through about 60 miles of atmosphere, so you don't see red light scattered across the sky when the Sun is at high noon.

At sunrise and sunset, as the Sun is at the horizon, light from the Sun has to travel through hundreds of miles of atmosphere to get to you. All that extra travel through the atmosphere gets the red part of sunlight to scatter. That's why the sky becomes red in the direction of sunrise and sunset. Baron Rayleigh figured that out. Smart cookie.

Baron Rayleigh had many other credits to his name, including the presidency of the Royal Society and Chancellor of Cambridge. His scientific discoveries in everything from the nature of sound to the properties of gasses proved invaluable. But his attempt to mathematically describe blackbody radiation was, well, catastrophic.

His partner in calamity, Sir James Jeans, made his name probing the properties of gasses. In addition to this, Jeans explored electricity and magnetism.

The two of them compiled research from the studies of blackbody devices. They analyzed the amounts of energy pumped into a blackbody device versus the intensities of the wavelengths of light that the blackbody device pumped back out.

Rayleigh and Jeans wanted to find an equation that would predict the amount of energy that a blackbody device (heated with a particular amount of energy) should radiate at the various wavelengths of the electromagnetic spectrum. They derived their equation using Newtonian physics, but something went very wrong.

The equation that they produced (the Rayleigh-Jeans law[114]) predicted how much energy would be radiated at which wavelengths when a blackbody device was heated.

But the Rayleigh-Jeans law went haywire at shorter wavelengths. While it sort of matched the experimental data at long wavelengths, at short wavelengths it went exponential, predicting that the blackbody device would radiate an infinite amount of energy off in the ultraviolet range.

That was obviously bogus. How could a blackbody device, which would seek equilibrium by dumping out the same amount of energy as was being pumped into it, produce an infinite amount of energy in the ultraviolet range?

This was certainly not what happened in the laboratory.

Since the Rayleigh-Jeans law went toward infinity at short, ultraviolet, wavelengths, this whole mess was named **The Ultraviolet Catastrophe**.

Now I don't know about you, but to *me,* the word "catastrophe" brings to mind images of massive loss of life and property. It makes physicists seem kind of wimpy for them to call the failure of one of their equations a catastrophe.

But maybe we should cut them some slack. For two centuries physicists had methodically grown a many-branched tree of knowledge from the seemingly indestructible roots of the Newtonian Laws of Nature.

To have the Rayleigh-Jeans law fail so spectacularly didn't just destroy one small branch of that tree of knowledge, it revealed that some of the basic assumptions of Newtonian physics were false. The entire tree was rotten at its roots.

Generations of scientists had devoted their lifetimes to decoding and expanding on what they thought were *the* Laws of Nature. Now it was revealed that the Newtonian tree of knowledge was unsound.

So yeah, I can see how this would seem catastrophic.

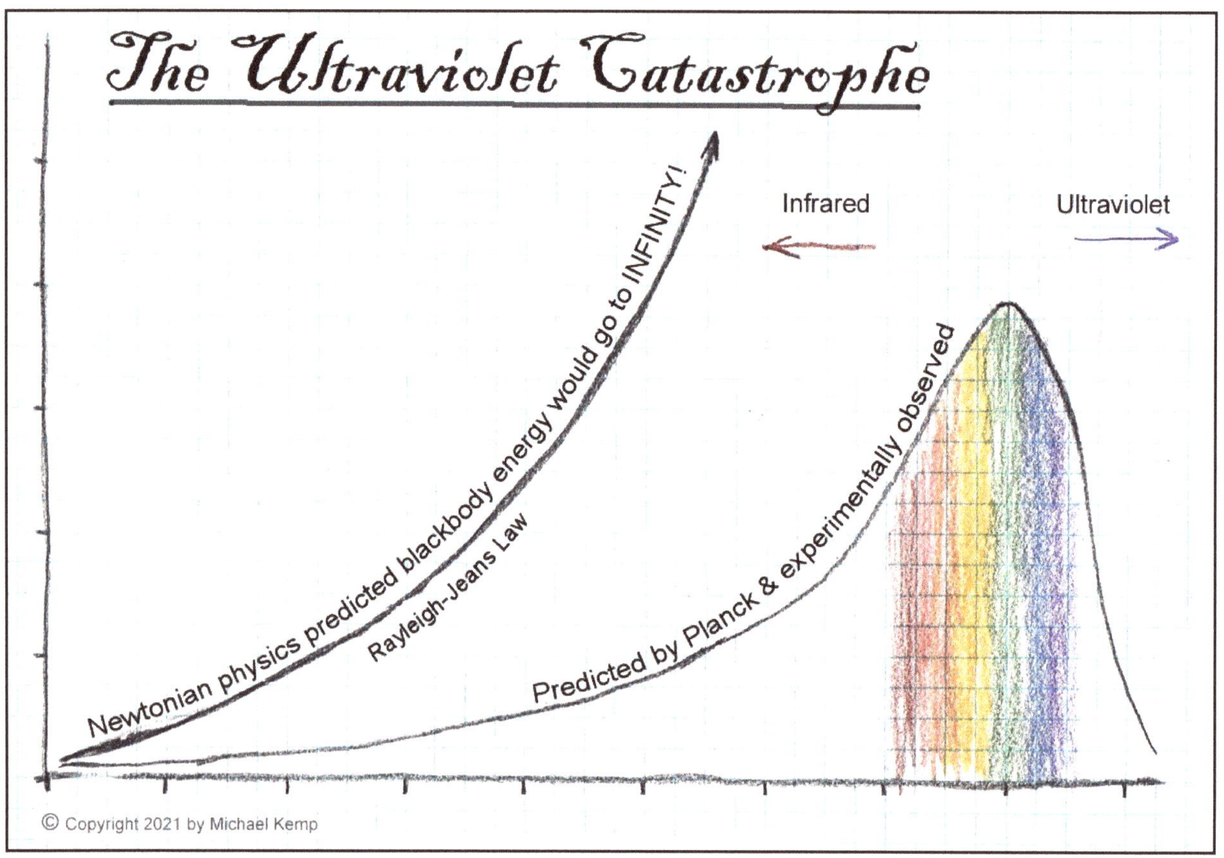

The Rayleigh-Jeans law describing blackbody radiation was faithfully derived from Newtonian physics, but the Rayleigh-Jeans law went haywire. So what was it missing?

Let me tell you a little about the character who resolved this catastrophe. Max Planck[115] grew up in a well-educated, traditionalist German family. Talented in many areas, he went into physics to satisfy his desire for order and settled knowledge. It was his focus on the study of heat and entropy that would eventually force Planck to "upset the applecart" and overturn what, up to that time, had been considered settled knowledge.

Beginning in 1894, Planck was working on formulas to explain the observed spread of energy across the spectrum produced by blackbody experiments. Just like Rayleigh and Jeans. He got very close, but his formulas never matched the experimental results at the shorter wavelengths. Just like Rayleigh and Jeans.

In the same way that Newton did not conceive of his theory of gravity in a simple "Eureka!" moment, Max Planck did not create his granular foundation of quantum mechanics overnight. He didn't even *want* to. But by 1900 he had realized that his Newtonian approach was never going to match the empirical data. In desperation, Planck turned to the works of Ludwig Boltzmann.[116] Boltzmann's work built connections between probability, entropy, and how a mind-boggling number of tiny events can combine to create what appears to be one single big, homogenous event. Boltzmann's ideas were not to Planck's liking — but he resigned himself to accepting Boltzmann's work. It was the only way that he could make his formulas work right and, incidentally, solve The Ultraviolet Catastrophe. He had to break the idea of the continuous nature of electromagnetic radiation down into a mind-boggling number of tiny individual particles of energy.

In Planck's own words, he embraced Boltzmann's work in *"an act of despair ... I was ready to sacrifice any of my previous convictions about physics."*

And that, my curious reader, is the true spirit of the scientific method. If the observed phenomena contradict what had been accepted as *the* valid description of nature (the time-tested theories, the Laws of Nature) — well then it's time to set the old theories on the shelf and start working up a new hypothesis that matches *all* of the facts, old and new.

Planck's new hypothesis was that light is created as individual particles. These particles come in fixed amounts of energy. It's as if energy could only be exchanged as one exchanges money. For example, the minimum currency in the United States is one penny. You do not have the ability to reach into your pocket or coin purse and pull out a smaller denomination of U.S. currency than one penny. You can spend multiple pennies, but not fractions of a penny. It turns out that our universe is built on a similar principle. Visible light and all electromagnetic waves come in particles that cannot be subdivided. They are uncuttable.

These days, we call these particles of electromagnetic radiation **photons**. According to Planck, the amount of energy contained in one of these photons is dependent on the wavelength of that photon. A long-wavelength photon has a small amount of energy. A short-wavelength photon has a large amount of energy. Some photons have a wavelength that is in the visible range of the electromagnetic spectrum. Some photons have a wavelength that is either too short or too long for our eyes to catch and process. Regardless of whether we can see them or not, electromagnetic radiation of any wavelength is "quantized" — electromagnetic radiation *always* comes in distinct particles — photons.

A blackbody device (or an iron poker left in the fire), when sufficiently heated, produces light. It produces photons, some of which have a wavelength that our eyes can catch and process. How does a blackbody device do that?

A blackbody device is built out of atoms. Each of those atoms are built with a central nucleus with a positive electrical charge. Electrons form the outer covering of an atom. Electrons have a negative electrical charge. Opposite electrical charges attract each other, and this is what keeps the electrons bound to the atom's nucleus. So far, so good.

When the blackbody device is heated, its atoms jostle against each other. That jostling knocks the atom's electrons into a higher-energy orbit around the atom's nucleus. But higher-energy orbits are unstable. To get back to a lower-energy orbit, an electron can bundle up the extra energy of that higher-energy orbit and create a photon out of it. Getting rid of that energy by creating a photon allows the electron to jump down to a lower-energy orbit. So when you see something hot, like an iron poker left in a fire, giving off red or orange light, that light is created by electrons jumping from higher-energy orbits to lower-energy orbits by packaging up their extra energy, using that energy to create photons, and sending those photons on their way. Each high-energy electron creates a single photon to carry off enough energy to let the electron jump to a lower-energy orbit.

Those photons carry off some of the heat energy, cooling the heated object. Those photons each carry a particular amount of energy, depending on their wavelength. The shorter the wavelength, the more energy one photon has.

When you heat a piece of metal in a fire, first it turns dull red. Red photons have the lowest frequency and longest wavelength that human eyes can catch and process. As the metal continues to heat up it goes from red to orange, to yellow, and finally to white-hot. At white-hot, the metal is actually creating photons over the entire range of visible light, red, orange, yellow, green, blue, and violet. When it gets to the point that it's putting out violet photons, it's also putting out much larger quantities of lower-energy photons, like red photons. Our eyes see this mishmash of many colors of photons as white, but all those colors are part of the white light, just like all the colors of the rainbow are hidden in sunlight until the sunlight goes through a prism.

When you heat a piece of metal, it starts giving off a dull red at around 900°F. It becomes cherry red around 1,400°F. It becomes bright yellow around 2,000°F. And by 3,000°F it shines white. So why doesn't a heated piece of metal put out a faint white light at 900°F, rather than giving off only red light? Here we have the crux of the matter.

Here we have Planck's solution to The Ultraviolet Catastrophe.

Planck proposed that light (and all electromagnetic radiation) is composed of individual particles. These individual particles (photons) are built with a fixed amount of energy. That fixed amount of energy is matched to their wavelength. The longer the wavelength and lower the frequency of the photon, the less energy that photon has. The shorter the wavelength and higher the frequency of the photon, the more energy that photon has. To create a photon of a particular wavelength, you have to have the energy needed for that wavelength. You can't create half of a high-energy photon. You have to have the full amount of energy needed for that wavelength.

A 900°F piece of metal produces a dark red light, but it cannot produce the orange, yellow, green, blue, or violet photons (all of which together make up white light) because these shorter-wavelength photons *require more energy* than the 900°F piece of metal has to offer. All it can offer is enough energy to send out longer-wavelength red photons. To produce a violet photon you need to have a lot more energy. To produce a higher energy photon, it's all or nothing. You can't produce half a violet photon, just like you can't hand the cashier at the grocery store half a penny. The 900°F piece of metal has enough energy to produce a humongous number of red photons but still does not have the energy to produce violet photons.

A blackbody device in the laboratory is in the same predicament. For a blackbody device at a particular temperature, there is a limit on how short of a wavelength of photons it can produce.

For a blackbody device that has a specific amount of energy being pumped into it, there will be a specific temperature that the blackbody device will rise to that allows it to get rid of as much energy as is being pumped into it. At that equilibrium temperature, there will be a general limit as to how much energy the individual atoms' electrons in the blackbody device are carrying. Because of the general limit to the energy that an individual electron will carry, these electrons will be limited as to the amount of energy that they can put into a photon.

Creating an ultraviolet photon requires a greater amount of energy than, say, creating an infrared photon. The consequence of this is that a blackbody device will be limited as to whether, and how many, high-energy ultraviolet photons it can produce.

Max Planck's proposal fixed the formula for calculating blackbody device radiation by putting a limit on how many photons can be produced at the shorter (ultraviolet) wavelengths. This limit predicted results that matched the experimental observations. This showed that the universe is indeed constructed in such a way that light is "quantized." Light comes in indivisible particles — individual photons.

They are minuscule, these photons. It takes a humongous number of them to add up to anything that we humans would notice in our daily lives. Things that look solid and continuous to us, such as the wavering flame of a candle, are sending out an *unimaginable* number of individual photons.

Oh why not! Let's try to imagine it anyway. I've seen estimates that a candle will put out anything from 10^{16} to 10^{20} photons per second. Let's start with 10^{16} to err on the conservative side. That's 10 million billion photons per second. To start to get a grip on this, think of grains of sand. We'll choose coarse sand that is about 1/16 of an inch in diameter. Think of looking out over a landscape of sand dunes made of this coarse sand. An area of those sand dunes that covers one square mile, and is one hundred feet deep would contain somewhere in the neighborhood of 10^{16} grains of sand. That is a *conservative* estimate of how many photons a candle releases each and every second. Imagine all that sand gone in one second. Each grain a photon.

If we went with 10^{20} photons per second, it would be more like an area of coarse sand three miles wide, six miles long, and a mile deep. Again, all gone in one second.

A candle is sending photons in all directions. When we look at the candle our eyes capture only a tiny fraction of the photons being sent out, but that is more than enough to make the wavering candle flame appear solid and beautiful.

E=hv

To express the relationship between a photon's wavelength and its amount of energy, Plank created a simple formula. In this formula he used frequency rather than wavelength, but you might recall that the frequency and wavelength are tied to each other. The higher the frequency, the shorter the wavelength. The lower the frequency, the longer the wavelength.

In my humble opinion, the equation **E=hv**, created by Max Planck, should be as well known as Einstein's famous **E=mc²** equation. Planck's **E=hv** equation solved The Ultraviolet Catastrophe. In doing so, it destroyed the continuous transitions of Newtonian physics and replaced them with granular bits of light and electromagnetism. This idea of granular bits of light and electromagnetism was the beginning of quantum mechanics.

In this equation **E=hv**:

- **E** stands for the energy of a single photon.
- **h** is called "Planck's constant." A *constant* in math jargon is a number with an unchanging value. For instance, the constant **π** (**pi**) is always 3.14159… I will go into what Planck's constant actually is shortly.
- **v** stands for the frequency of the photon's vibration.

What **E=hv** is trying to say is: if you want to know how much energy a single photon has, you multiply its frequency by Planck's constant to find out.

Planck's constant "**h**" is an unchanging number, so the only thing that causes a difference in the amount of energy in one photon compared to another photon is the difference in those photon's frequencies. The higher the frequency, the bigger that number **v** is going to be.

E=hv is math-speak for the idea that higher frequency photons have more energy in each photon, and that lower frequency photons have less energy in each photon.

Ultraviolet photons, with short wavelength and high frequency, have a large amount of energy in each individual photon.

Infrared photons, with long wavelength and low frequency, have a small amount of energy in each individual photon.

The limitation on how many high-frequency ultraviolet photons a heated blackbody device can create corrects the Rayleigh-Jeans Law's Ultraviolet Catastrophe. This was the fix needed to make the physicist's equations match the experimental results.

In the equation **E=hv**, the **"h"** is referred to as Planck's constant. He did not just make it up, he had to derive its value from previously verified values for how electromagnetic waves and energy actually work. **h** serves to translate between the human unit of measure for energy and the human unit of measure for frequency. In his equation **E=hv**, the **E** for energy is measured in units called Joules, which are based on the human-defined measurements for force, distance, and time. The **v** represents the photon's frequency measured in cycles per second, a second being the human-defined measurement of time.

Apparently, modern humans were not consulted when the universe was being formed, and so our units of measurement need correcting factors to come into line with how the universe is actually built.

In the equation **E=hv**, this correcting factor is Planck's Constant: **h**. The value of **h** turned out to be approximately 6.626×10^{-34} $meters^2 \times kilograms$ / second, which seems pretty random and crazy, but it bridges the gap between our human units of measure and how our universe is actually constructed.

We get so used to using human-defined units of measure, that we think that they are especially meaningful. Length in meters or feet. Weight in pounds or kilograms. Time in seconds. But each of these are based on tradition, legend, or on our physical surroundings here on Earth.

Take time, for instance. A year is the amount of time it takes the Earth to orbit once around the Sun. A day is the amount of time it takes the Earth to rotate once on its axis. Those are based on our unique planet, not on the structure of the universe at large. Our finer measurements of time are based on human preferences for specific numbers that were considered special in some way. These smaller units of time were set in place by the ancient Babylonians, Egyptians, and Greeks. Follow this link[117] if you'd like to see the details.

We have made up our units for measuring length, weight, and temperature to suit our fancy. In the metric system these are based on the distance from the Earth's equator to its North Pole, and on the attributes of pure water when it is at sea level. In the American system, these units of measure are based on traditions and legends.

Another form of Planck's Constant is written as **ħ** with a little horizontal dash through the top of the **h**. "**ħ**" is shorthand for the value **h/(2π)**. It turns out that a number of useful physics equations use the term "Planck's Constant divided by two times Pi," so it became shorthand to write that as **ħ**. This is referred to as the "reduced Planck's constant." You will see the reduced Planck's constant pop up in Appendix B, in another equation that shook physics.

This wraps up our journey to understand why and how scientists were forced by experimental results in the 1800s to recognize that not only is matter made up of tiny bits, such as atoms and electrons, but even light is built up from an unimaginable number of tiny bits. Quantum particles. Photons. Planck's work was a turning point, leading scientists to propose, test, and confirm that all of our physical reality is built up from an unimaginable number of quantum particles of matter and force.

In Appendix B, we will come at quantum mechanics from another angle, and follow how quantum mechanics not only describes the universe as composed of quantum particles, but also the absurd behavior of those quantum particles.

Review

William Hershel discovered that different colors of sunlight transmit different amounts of heat. Experimenters followed this up by artificially generating light by pumping energy into blackbody devices until they created light and then measuring the amount of energy being dumped out of the blackbody device at various wavelengths of light.

The trouble was that when Baron Rayleigh and James Jeans tried to use Newtonian physics to predict this output of energy across the electromagnetic spectrum, their equations predicted infinite energy output in the ultraviolet end of the spectrum. This was ridiculous. More to the point, it was not what the experimenters had observed. This failure of the Rayleigh-Jeans Law created what is called "The Ultraviolet Catastrophe." It showed that there are ways in which the universe does not play by the assumptions of Newtonian physics.

The proposal that Max Planck put forward to solve The Ultraviolet Catastrophe revealed that, at an infinitesimally small scale of existence, our universe is granular, not continuous and solid. Electromagnetic energy, including visible light, comes in individual particles — photons. A crucial aspect of Planck's discovery was that the shorter the electromagnetic wavelength, the more energy an individual photon has. The longer the wavelength, the lower the energy per photon. Planck's equations matched experimental results, but Planck's equation broke the Newtonian belief in continuously variable forces of nature to bits.

In the same way that Steven and de Groot disproved Aristotle's gravity, Planck disrupted the foundation of Newtonian physics.

The veil of conventional belief had once again been parted to reveal the stranger nature of our universe. Stay tuned — Appendix B is where this stuff gets truly bizarre.

Activity

How Many K Is Your Light Bulb?

I'll give you a real do-it-at-home quantum mechanics activity at the end of Appendix B, but here's something to tide you over.

If you have a spare package of light bulbs in the cupboard, somewhere on the package there is usually a "color temperature" chart. If you don't have a spare package at home, you can look up light bulbs online. The color temperature chart is usually a bar going from 2500K to 5000K with a marker showing where these particular light bulbs are within that range. The "K" stands for Kelvin. Kelvin is the temperature scale that starts with its "zero" degrees set to absolute zero (the theoretical coldest temperature that could ever be possible).

We saw how light gets created when a blackbody device gets heated up. And as the blackbody device gets hotter, the light goes from reddish to white, or even bluish. Hotter blackbody light has more blue and violet in it. The less heat the blackbody has, the less blue and violet is in the mix and the redder the light looks.

The higher the "K" in the light bulb's color temperature chart, the more blue light you get from that light bulb.

You might notice something backwards about the color temperature chart. The left end of the chart has the lowest Kelvin temperature, and yet it is often labeled "Warm." The right end of the chart has the highest Kelvin temperature, and yet it is often labeled "Cool." What's going on here? Why is a lower temperature labeled Warm while a higher temperature is labeled Cool?

In our daily life, things that are *very* hot may give off a red light, like the coals in a fire. On the other hand, when we look at photos of the cracks in a glacier we notice a bluish light. A lake of cool water looks bluish. Common sense tells us that red is hot and blue is cold.

But that's not how individual photons work according to Plank's formula. A blue photon has more energy (and thus a higher Kelvin temperature rating), and a red photon has less energy (and thus a lower Kelvin temperature rating).

So right there on your lightbulb package you can see Planck's formula playing out in the red-to-blue Kelvin rating. Lower energy, redder wavelength. Higher energy, bluer wavelength.

At the link below, you can play with how color temperature (K) changes the energy/wavelength profile of light. This is like those graphs I made of the energy/wavelength of light that the Sun puts out, and for The Ultraviolet Catastrophe — except that my graphs put long wavelength (low energy) light on the left side, while this link puts short wavelength (high energy) light on the left side. When you change the Blackbody Temperature slider on the right side of the graph at this website, you can see the energy/wavelength profile change. Also note how the B[lue] G[reen] R[ed] indicators and the starburst icon above the energy/wavelength graph change as you slide the Blackbody Temperature up and down.

https://phet.colorado.edu/sims/html/blackbody-spectrum/latest/blackbody-spectrum_en.html

Links

Biography
108. "William Herschel | Biography, Education, Telescopes, & Facts ...." 11 Nov. 2020, https://www.britannica.com/biography/William-Herschel.

Deeper Dive
109. "Herschel's Experiment | Cool Cosmos."
https://coolcosmos.ipac.caltech.edu/page/herschel_experiment.

Deeper Dive
110. "Black-body radiation - Wikipedia."
https://en.wikipedia.org/wiki/black-body_radiation.

Biography
111. "Gustav Kirchhoff - Biography, Facts and Pictures."
https://www.famousscientists.org/gustav-kirchoff/.

Biography
112. "John William Strutt, 3rd Baron Rayleigh - Wikipedia."
https://en.wikipedia.org/wiki/John_William_Strutt,_3rd_Baron_Rayleigh.

Biography
113. "Sir James Jeans | British physicist and mathematician | Britannica."
https://www.britannica.com/biography/James-Jeans.

Deeper Dive/Physics-Speak
114. "Rayleigh–Jeans law - Wikipedia."
https://en.wikipedia.org/wiki/Rayleigh%E2%80%93Jeans_law.

Biography
115. "Max Planck - Wikipedia."
https://en.wikipedia.org/wiki/Max_Planck.

Biography
116. "Ludwig Boltzmann - Wikipedia."
https://en.wikipedia.org/wiki/Ludwig_Boltzmann.

Deeper Dive
117. "Why is a minute divided into 60 seconds, an hour into 60 minutes." 5 Mar. 2007,
https://www.scientificamerican.com/article/experts-time-division-days-hours-minutes/.

Deeper Dive
118. "What is absolute zero?"
https://www.nbcnews.com/mach/science/what-absolute-zero-ncna936581

Appendix B:

Unveiling the Absurd (Quantum Mechanics Part 2)

"Quantum mechanics describes nature as absurd
from the point of view of common sense.
And it fully agrees with experiment.
So I hope you can accept nature
as She is — absurd."
~ Richard Feynman

In Appendix A, we saw how the study of heat and light revealed that electromagnetic radiation (including visible light) comes in distinct quantum particles. Photons. In physics, light is considered to be a force, so a photon is a quantum particle of force.

In the late 1800s, English physicist J.J. Thomson demonstrated the existence of a quantum particle of matter called the electron. So by 1900 we come to a point in history where it was recognized that both matter and force come in the form of quantum particles.

In Appendix B, we will look at another line of inquiry that started in the 1800s. This series of experiments revealed the *behavior* of quantum particles. Appendix B follows the dawning realization that quantum particles do not *behave* as you would expect a "particle" to behave. Most of the time quantum particles behave more like little waves. It is only when a quantum particle interacts with its surroundings that it behaves like we would *expect* a tiny particle to behave.

Let's look at the real-world observations which revealed the absurd behavior of quantum particles: wave-particle duality; the probabilistic nature of the quantum realm; the observation that quantum particles "jump" from one state to another; and of course Heisenberg's uncertainty principle.

These strange behaviors have nothing to do with the difficulty in observing such tiny bits of matter and force. These strange behaviors are the basic nature of quantum particles. As much as common sense may rebel, these behaviors were revealed by real-world observations.

Let's follow this trail of discoveries.

Light Travels Like a Wave

The first experiment in the 1800s to lay groundwork for understanding the strange behavior of quantum particles, was a demonstration that light travels like a wave. This was in conflict with the popular idea that light was a particle.

The influential French philosopher René Descartes and the respected scientist Isaac Newton maintained in the 1600s and 1700s that light was composed of tiny bits of matter, which they called corpuscles.[119]

Then, in 1801, a simple experiment revealed **interference patterns** in light. Only waves make interference patterns. So this experiment showed that, when it is traveling, light behaves like a wave. This discovery was so at odds with the accepted corpuscle theory that it took years before it was accepted in the general scientific community. In fact, it took a man with established credibility in the scientific community, named Augustin-Jean Fresnel (the inventor of the Fresnel lens, which is used in lighthouses), to champion the results of this experiment, for it to finally gain acceptance.

But first a little background. What are these "interference patterns" that I mentioned? An interference pattern is what happens when two waves interact.

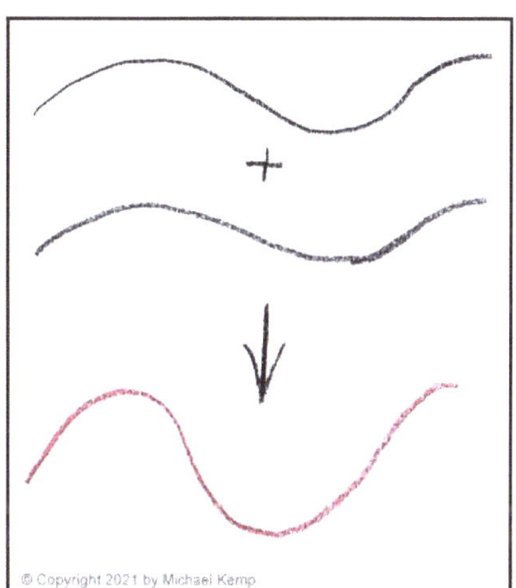

In the illustration on the right, the first wave (top) and the second wave (in the middle) have their peaks and valleys lined up, and if they interact with each other they produce a *larger* wave (bottom). This is called *constructive* interference.

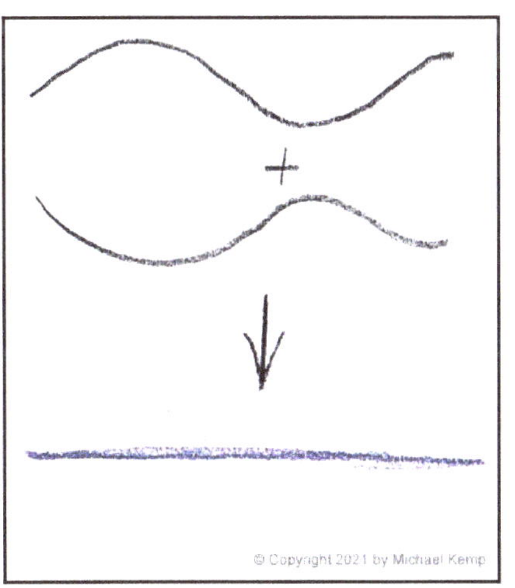

In the illustration on the left, the first wave's peaks line up exactly with the second wave's valleys. If these waves interact with each other they will *cancel* each other out, leaving *no wave at all* (bottom). This is called *destructive* interference.

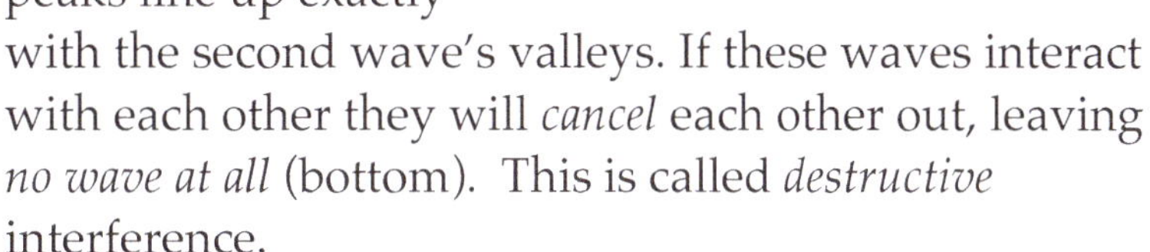

There is a crucial condition for producing a recognizable interference *pattern*. It is called **coherence**. Coherent light is light that has a single wavelength. Another way to say it is that it is light of a single color. All of the waves of coherent light have the same distance between their peaks and troughs, the same wavelength, the same frequency. This can create repeating patterns of interference that are easy for us to see.

If you live near a pond or lake, you can demonstrate interference patterns for yourself with a couple of pebbles. Try this as an extra-credit activity, or as a thought experiment. If you drop two similar pebbles, separated by a few inches, into still water, each pebble will create a series of expanding circular waves. And these two sets of waves interfere with each other. This interference creates its own pattern superimposed on the circular waves that the individual pebbles created.

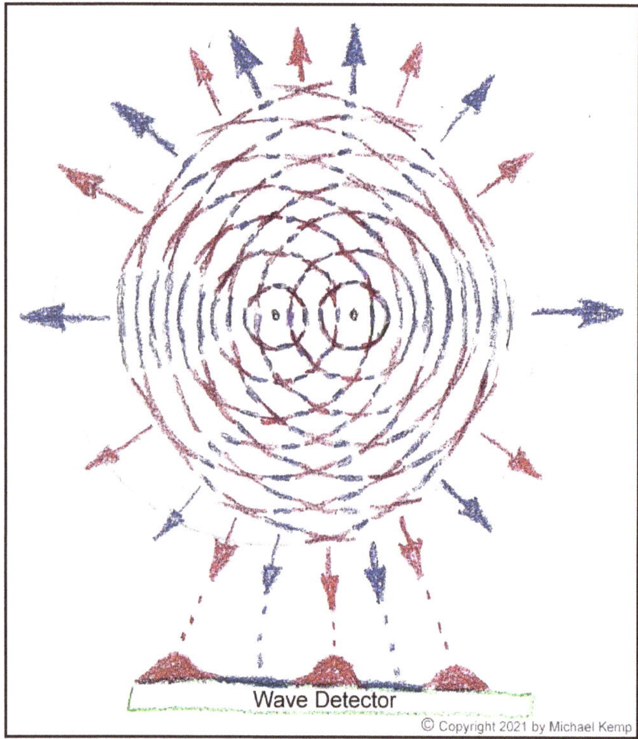

Here is a drawing of how it looks from above when the two pebbles are dropped in still water.

The red parts of the ripples (and red arrows) show where the two ripples reinforce each other, creating higher peaks and lower valleys. In this illustration, the blue parts of the ripples (and blue arrows) are where a peak from one pebble's ripple meets the valley of the other pebble's ripple, causing the two ripples to cancel each other out. That creates rays of still water.

If some sort of "wave detector" is placed as shown at the bottom of the illustration, the red lumps show where the waves hitting the detector are highest and the blue lines show where waves at the detector are canceled out.

The Double-Slit Experiment

Now we are ready to dig into the experiment that showed that light travels like a wave.

Meet the British scientist, Thomas Young, who made contributions to our understanding of energy, light, and vision — as well as language, music, and archeology. In 1801 he performed an experiment that established, beyond question, that light travels like a wave.[120]

In Young's 1801 double-slit experiment, he used diffracted sunlight to select a single coherent wavelength of light. He then directed this coherent light onto a screen that had two slits. This created two new radiant sources of coherent light coming out the other side of the screen. The light from these two slits shone on another solid screen (like the "wave detector" in my example of dropping two pebbles in a pond).

An interference pattern appeared: repeating bands of light and dark were seen. The waves of coherent light from the two slits were interfering with each other, in some places reinforcing each other and in other places canceling each other out. Just like you would see from the waves in water when two pebbles are dropped near each other.

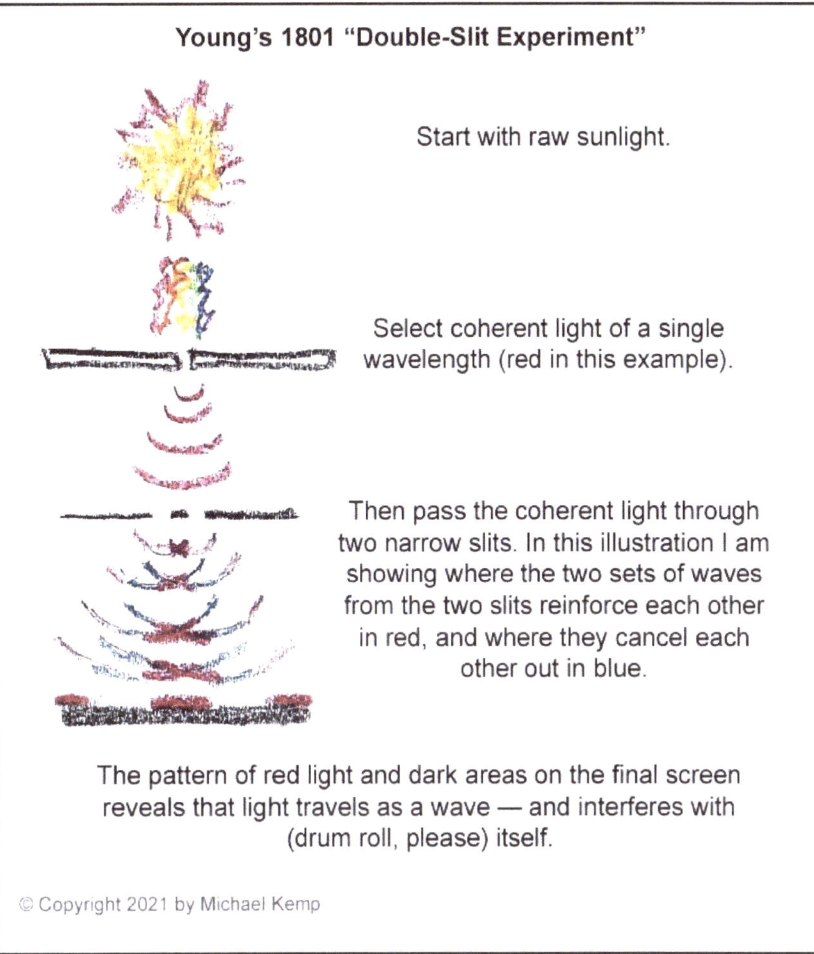

I'll leave you with one more bit of strangeness to ponder about the "double-slit experiment." This type of experiment has been done where only one photon of light is sent through the double-slit at a time. As individual photons are recorded on the detector screen, each photon leaves its mark at a specific spot on the detector screen. As more photons are sent through the experiment, they gradually add up to the bars expected in an interference pattern. Each photon's wave travels through both slits and *interferes with itself,* determining where it should, or shouldn't, land on the detector screen. Each photon is first a *wave* that goes through both slits and *interferes with itself,* but then it interacts as a *particle* that is recorded at a single point on the detector screen.

So is a photon a wave, or is it a particle?

Yes.

If that makes your brain melt, don't worry, it makes *everybody's* brain melt the first time they run into this wave-particle duality. And yet, you can't dismiss it, because this is how light actually behaves in our universe.

Atomic Fingerprints in the Rainbow

Hydrogen is the simplest atom in the universe. In its center is a single proton which has a positive electrical charge. This proton is a hydrogen atom's nucleus. A hydrogen atom also has a single negatively charged electron that "orbits" the nucleus. That's it. One proton, one electron. The simplicity of the hydrogen atom makes it a great test subject for studying how atoms are built.

In the mid-1800s, the Swedish physicist Anders Jonas Ångström[121] performed some curious experiments with pure hydrogen gas.

By "pure hydrogen gas" I mean that a container (for example, a glass tube) would be filled with nothing but hydrogen atoms.

In one of Ångström's experiments he energized this hydrogen gas to the point that it produced light, much like a neon sign. This can be accomplished by passing electricity through the hydrogen gas, in the same way that a neon sign is energized to make it produce light.

Ångström used prisms and diffraction grating to spread light out into all the colors of the rainbow. But when he spread out the light from his energized hydrogen into a rainbow, it had only four lines of color. There was a red line, a bluish-green line, a bluish-violet line, and a faint deep-violet line. The rest of the rainbow was dark. The energized hydrogen gas was producing light at only four distinct wavelengths in the range of visible light. This was the hydrogen atom's fingerprint in the rainbow.

Then Ångström tried something else. He started with white light, and when you pass *that* light through a prism it produces the full rainbow, without any gaps in the colors of the rainbow.

He took a container of pure hydrogen gas that was *not* energized, and passed this white light through the cold hydrogen gas. You could say that the white light was *filtered* through the cold hydrogen gas. Then he spread that cold-hydrogen-filtered light out into a rainbow. This rainbow had four dark lines at the same four wavelengths where the energized hydrogen gas had shown four bright lines.

Cold hydrogen was sucking up the same wavelengths of light that energized hydrogen gas produced. The cold-hydrogen-filtered light was like a negative fingerprint in the rainbow.

Hydrogen is not unique in this behavior. Each atom has its own fingerprint in the rainbow. If you hear someone talk about "spectroscopy" or "spectral analysis," hydrogen's fingerprint in the rainbow is the simplest example of what they are talking about.

It seemed curious, it was a clue of some sort, that the simple hydrogen atom only produced or sucked up visible light at these four distinct wavelengths. Why would hydrogen have this fingerprint in the rainbow?

If you read science articles, you will see that the bright lines produced by a heated gas are called *emission* lines. The dark lines in cold-filtered-light are called *absorption* lines. You often see these mentioned when folks are talking about the light from distant stars and galaxies. A star is made up of extremely energized gas, and its emission lines tell us which gasses are in that star. On the other hand, when distant starlight passes through a cold interstellar dust cloud, and that filtered starlight is spread out into a rainbow, the dark absorption lines in that rainbow tell us which gasses are in that cold dust cloud.

It took until the early 1900s, before these fingerprints in the rainbow finally started to make sense. We will dig into the discoveries that revealed the strange quantum behaviors which produced these fingerprints in the rainbow in the next few sections. These discoveries, along with Planck's quantized light, are the cornerstones of quantum mechanics.

Leaping Electrons

Heinrich Hertz was born in 1857 in the sovereign state of Hamburg within the German Confederation. He studied science and engineering in Dresden, Munich, and Berlin, eventually becoming a full professor in the town of Karlsruhe in southwest Germany.

His obsession was to experimentally verify the predictions of James Clerk Maxwell's electromagnetic theory (see *Electric Light* in Chapter 8). In 1887 he set up an experiment to demonstrate that when he created a powerful spark at one end of a room, it would produce electromagnetic waves which would induce a current in a loop of wire at the other end of the room. That loop of wire, which received the electromagnetic waves, was an antenna. Hertz would detect the current in the antenna by putting the end of a separate grounded wire very close to the antenna, and watching for a spark from the antenna to the grounded wire.

One of Hertz's students, Philipp Lenard, demonstrated that the electric sparks seen by Hertz were composed of electrically charged particles of matter: what we now call "electrons."

Hertz noticed something strange.

When Hertz put a pane of glass between the big spark emitting the electromagnetic waves, and the antenna, there would be no spark between the antenna and the ground wire. Glass lets visible light through, but blocks ultraviolet light. A sheet of quartz lets *both* visible light and ultraviolet light through. When he tried using a sheet of quartz rather than glass, then the spark between the antenna and the ground wire came back. It was only ultraviolet light, not visible light, that would excite the antenna to the point that an electrical spark would jump off the antenna to the ground wire. This curious relation between ultraviolet light and electricity became known as the **photoelectric effect**.[122] [123]

Hertz continued to focus on other predictions of Maxwell's equations, but he wrote that *"…the effect [of ultraviolet light] is striking and yet totally puzzling. Naturally it would be nicer if it were less puzzling; however, there is some hope that, when this puzzle is solved, more new facts will be clarified than if it were easy to solve."*

Years later, Einstein took up the challenge.

The puzzle was that this dependance on *ultraviolet* light did not match Newtonian physics. According to Newtonian physics, when light was focused on metal, the energy of the light would cause electricity to radiate off of the metal. So far, so good. This Newtonian version of the photoelectric effect said that it should be continuously variable. That is to say, a little light should produce a small amount of electricity. A large amount of light should produce a larger amount of electricity. It shouldn't be tied to the wavelength of the light.

That is not what was observed in experiments. In real-world experiments, no electricity was produced at the longer wavelengths of light, no matter how bright that light was. Electricity was only produced when the wavelength of the light was tuned to the ultraviolet range.

Einstein took the experimental result that electricity was only produced when the light was tuned to ultraviolet wavelengths, plus some current speculations about the structure of matter, plus Planck's recent revelations about the quantum nature of photons, and proposed that the photoelectric effect was quantum, rather than Newtonian.

The quantum nature of the photoelectric effect could be explained:

- If the electricity coming off the antenna was composed of tiny bits of electrically charged matter: electrons. This idea that electricity was composed of electrons had been put forward by others in the 1800s.
- If matter was made out of atoms which contained those electrons. It had recently been proposed that each atom was constructed from an electrically positive blob,

with enough negatively charged electrons mixed in to make the atom as a whole electrically neutral. This was J.J. Thomson's "plum pudding" model.

- And if light was quantized as Planck had said, and the energy of a single photon was tied to its wavelength by Planck's **E=hv** formula.
- And if the electrons in the antenna, or more specifically, the electrons in the atoms that the antenna was made out of, were bound to their atoms by a "quantized" amount of energy. An amount of energy equal to an ultraviolet photon.
- And if individual photons interacted only with individual electrons.
- And if an electron would not store up the energy of multiple low-energy photons. It would only be ejected from the metal wire of the antenna if the incoming photon had enough energy to eject it. One and done, or the electron would completely ignore the photon.

This introduced the idea of quantum energy levels in how electrons were bound to their atoms. A quantum particle of matter (the electron) would interact with a quantum particle of force (the photon) only if the photon had an energy level that the electron could immediately use. If the photon's energy level was not something that the electron could immediately use, the electron would not interact with the photon. If the photon's energy level *was* something that the electron could immediately use, the electron would absorb and destroy the photon, and immediately make use of that energy. In the case of the photoelectric effect, the electron would use the photon's energy to jump off the antenna to the ground wire.

An Atom Isn't Built The Way It "Should" Be

The ancient Greeks conceived of atoms as tiny, uncuttable particles.

By the early 1900s it was recognized that atoms had some sort of internal structure.

Ernest Rutherford was a New Zealand physicist, and world leader in researching radiation, atoms, and subatomic particles. It was his experiments bombarding thin gold foil with "alpha particles" (bundles of two protons and two neutrons each), and observing how those alpha particles were deflected, that led him to propose a new model for the atom in 1911.

The drawings of atoms that you generally see today are based on this "Rutherford model." My drawing (left) of a lithium atom is an example. There is a small central **nucleus**, composed of **protons** (having a positive electrical charge) and **neutrons** (having no electrical charge). Orbiting this nucleus are negatively charged **electrons**.

The Rutherford model of an atom is used in the logo of the Atomic Energy Commission, and in innumerable science-oriented logos and graphics.

The Rutherford model of an atom is not quite right.

The idea of a tiny, dense nucleus surrounded by a larger cloud of electrons is correct. But there's something wrong with the idea that the light-weight electrons orbit the massive nucleus just like the Moon orbits the Earth. For one thing, the electrons are not bound to the nucleus by gravity. Gravity is too weak at this scale. Electrons are bound to the nucleus because their negative electrical charge is strongly attracted to the positive electrical charge of the nucleus. But that's not the *real* problem with Rutherford's atomic model.

Quantum effects make the electrons' "orbits" around the nucleus much stranger than little particles going in circles!

We have to go through a couple of layers of quantum effects to get to a model of atomic structure that matches how atoms actually act in the real world. The first layer of quantum strangeness explains why each type of atom has a "fingerprint in the rainbow." The second layer of quantum strangeness goes beyond "strange," into the bizarre world of quantum waveforms.

Quantum Orbits

Niels Bohr[124] was a Danish physicist, with keen interests in mathematics and philosophy. He played several key roles in the development of quantum mechanics. He pioneered a great deal of early quantum research and theory. He also acted as a mentor for those entering this new field of science.

In 1921, Bohr's dream of an institute of physics at the University of Copenhagen was realized, with Bohr as the director. Most of the bright young minds that hammered out the details of modern quantum mechanics spent time in Bohr's "Copenhagen School."

If you see someone use the phrase "the Copenhagen School" of quantum mechanics, or "the Copenhagen interpretation" of quantum mechanics, Bohr's institute is what they are referring to.

But let's get back to making a better model for the structure of the atom. In 1913 Bohr proposed a new model of the atom[125] that:

- Fixed a problem with the Rutherford model.

- Introduced the quantum energy levels of electrons that Einstein had proposed in his 1905 paper on the photoelectric effect.
- Explained Ångström's discovery that atoms have a fingerprint in the rainbow.

All Atoms Would Become Inert Lumps

The *real* problem with the Rutherford atomic model is that if electrons were truly running circles around the atomic nucleus, they would lose energy and crash into the nucleus.

Let me explain.

First off, in the experiment that Hertz did (described in the *Leaping Electrons* section of this appendix) he created electromagnetic radiation by generating a large spark. That spark was a stream of electrons being accelerated across an air gap. One of the lessons learned from this was that **an accelerated electron produces electromagnetic radiation**. This electromagnetic radiation carries off energy from the electron.

Secondly, as we saw in both Newton and Einstein's description of a circular orbit (from *An Orbit Is Gravity Balanced by Inertia* in Chapter 5, and *Einstein's Orbit* in Chapter 10) an object in orbit is **constantly being accelerated** toward the object that it is orbiting around.

It follows that an electron in a circular orbit (being constantly accelerated) will be constantly producing electromagnetic radiation (photons). Those photons will constantly be carrying off energy from the orbiting electron. As the electron loses energy, it will not be moving forward fast enough to resist the pull of the nucleus. It will spiral into the nucleus, radiating photons all the way down.

This would cause each and every atom in the universe to decay into an inert lump in

If electrons orbited the atom's nucleus as shown in the classic "atom" drawing, then by all that we know about charged particles, the electron would continuously be producing photons. The photons would drain energy from the electron's orbit, causing it to spiral into the atom's nucleus in the blink of an eye. If electrons really orbited the atomic nucleus like this, then all of the atoms in the universe would implode in a flash of photons.

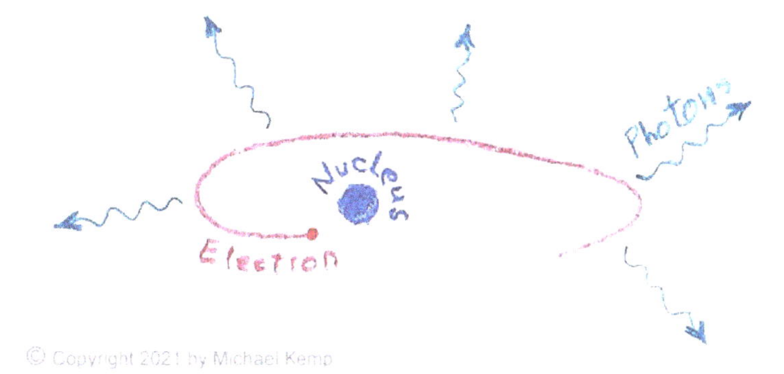

the blink of an eye. Everything held together by atomic bonds would disintegrate. Trees, buildings, you, me, ice cream, and lip gloss. All gone in a flash of photons.

Since that has not happened, the Rutherford model cannot be the whole story.

Quantum Energy Levels to the Rescue!

In Einstein's 1905 paper on the photoelectric effect, he proposed that electrons bound to the atoms of Hertz's antenna were bound at quantized energy states. If the incoming photon did not have the right energy to break that quantized bond, the electron would not absorb the photon.

Bohr expanded on that idea. Bohr's description of the atom, published in 1913, assumed that the energy levels of the electron orbits around the nucleus were quantized. The quantum energy levels of these orbits would be based on the mass and inertia (in the form of angular momentum) of the electron. These quantum energy levels, and the resulting size of the orbits could be calculated from the reduced Planck's constant (**ħ)** multiplied by a positive integer (1, 2, 3, …). Bohr's model of the atom is a bit more complicated, but this gives you the basic idea.

In Bohr's model of the atom, electrons could exist *only* in one of these quantum orbits. An electron in a quantum orbit would not produce photons, would not lose energy, and would not spiral down into the nucleus. The electron would *only* produce or absorb a photon by jumping between the allowed quantum orbits.

To jump from a higher energy orbit to a lower energy orbit, the electron would create a photon of the exact energy difference between those orbits. It would send the photon off on its way, and jump to the lower energy orbit.

The only way an electron could jump from a lower energy orbit to a higher energy orbit would be if it could absorb and destroy a photon, and immediately use the energy of that photon to make the jump to the higher energy orbit. The electron would do this *only* if the incoming photon had exactly the right energy to match the difference between the electron's current orbit, and a higher energy orbit.

Either jumping up or jumping down between quantum orbits of different energy levels, the photon involved had to match the difference in those energy levels exactly.

Bohr's Model of the Atom Made Unique Predictions

Bohr's model made some unique predictions.

Bohr's model described the radius of the various quantum orbits for the hydrogen atom. The measured size of hydrogen atoms matched Bohr's prediction of the radius of hydrogen's minimal energy level orbit. Not bad.

The energy levels of the quantum orbits, in turn, define the energy *difference* between those quantum orbits. According to Planck's formula **E=hv**, a photon of a specific energy has a specific wavelength. Specific energy differences between quantum orbits dictate the energies, and therefore the wavelengths, of the photons that can allow an electron to jump between the quantum orbits of the hydrogen atom.

These photon wavelengths matched the lines in the rainbow that Ångström observed. The bright lines from Ångström's energized hydrogen matched the photons that Bohr's model said would be produced by electrons shedding energy in order to drop to a lower energy orbit.

Likewise, the dark lines from Ångström's cold-hydrogen-filtered light matched the energies of photons that would be absorbed and destroyed by electrons as they jumped to a higher energy orbit. Hydrogen's fingerprint in the rainbow was now explained. Not bad at all.

The following oversimplified illustration is of Bohr's model of the hydrogen atom. Each circle represents one of the quantum orbits.

Hydrogen has only one electron, but I've drawn two versions of that single electron: once for absorbing a photon and jumping up to a higher energy orbit, and again for producing a photon with enough energy to allow it to drop down to a lower energy orbit.

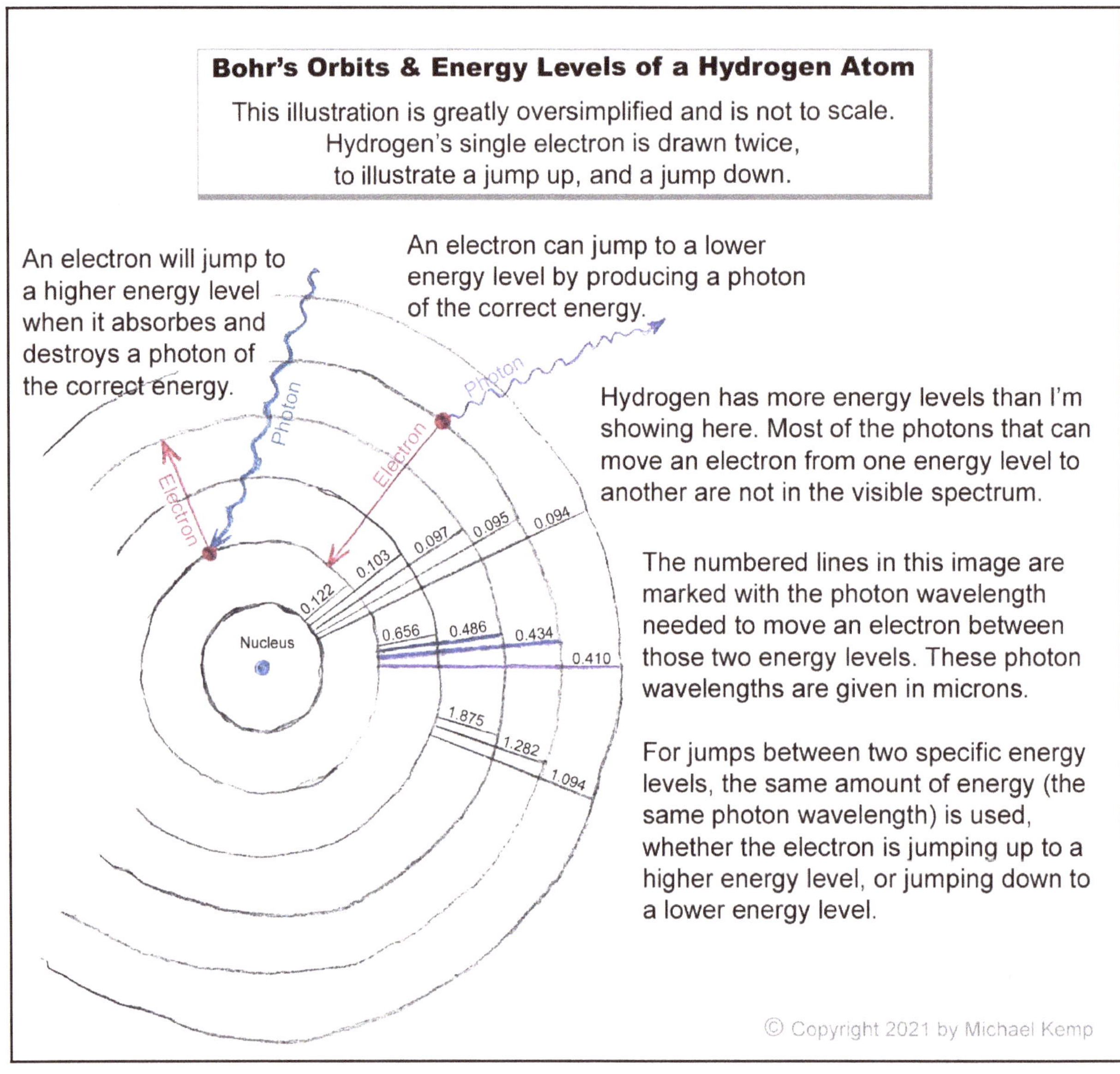

Bohr's atomic model not only matched the characteristics of the atoms that were then known, it also predicted the characteristics of as-yet-undiscovered atoms. When the first of these atoms *was* discovered and experimental testing showed that it had the characteristics predicted by his atomic model, it showed just how powerful Bohr's model really was.

The Absurd Behavior of Quantum Particles

Are you sitting down? As strange as Bohr's quantum orbits are, we need to go through another layer of quantum behavior to get to a model of the atom that matches how atoms act in the real world. It's about to go from "That's strange" to "That's absurd." But, to repeat the quote at the beginning of this appendix:

"Quantum mechanics describes nature as absurd
from the point of view of common sense.
And it fully agrees with experiment.
So I hope you can accept nature
as She is — absurd."
~ Richard Feynman

Wave-Particle Duality

As noted at the beginning of this appendix, the double-slit experiment done in 1801 by Thomas Young showed beyond a doubt that light travels like a wave.

From Einstein's description of the photoelectric effect and Bohr's model of the atom, a photon interacts with an electron as a distinct bundle of energy — as a quantum particle. So, photons travel like a wave but interact like a particle. This is the quantum wave-particle duality.

In physics-speak, the most elementary quantum particles are either "matter" or "force" particles. Photons are electromagnetic quantum particles of force.

In the 1920s it was slowly sinking in that the wave-particle duality was not limited to photons, or even to quantum particles of force. In what is known as the Davisson–Germer experiment (performed from 1923 to 1927) they fired beams of electrons, which were scattered by the surface of a nickel crystal. The scattered electrons created diffraction patterns. Like an interference pattern, a diffraction pattern is only observed in waves. If you want to know more about the Davisson-Germer experiment and are fluent in physics-speak, just follow this link.[126] These experimental results showed that an electron also travels like a wave. The wave-particle duality extends to all quantum particles, regardless of whether they are matter or force. Quantum particles are sometimes called "wavicles," in recognition of this wave-particle duality.

"If matter acts like a wave" I hear you asking, "then why don't we see that play out in our everyday lives?" This is a valid question. The short answer is "It's a matter of scale."

Quantum wave functions operate at an incredibly tiny scale. At the human scale of existence, even the smallest objects that we interact with are composed of such a humongous number of quantum particles that their wave nature is totally blurred out. That goes for even a speck of dust.

You could use the analogy of a high-resolution photograph. If you use a microscope on such a photo, you can see the individual pixels that it is composed of — but to the naked eye, those pixels blur into a smooth image. In the same way, the wave nature of quantum particles is completely blurred out at the human scale of existence.

Quantum Waves Ride on a Quantum Field

To know that experiment shows that quantum particles travel like a wave is all well and good, but what is this wave?

One way to think about this "wave function" is that it represents the *probability* that the quantum particle will have specific values at specific points in spacetime. What are the chances that it will be *here* in another nanosecond, and what are the chances that it will be *there*?

Another way to think of this quantum "wave" is that it is a very local disturbance rippling across a quantum field of existence.

Let's set the stage, and see how these ideas progressed.

Werner Heisenberg[127] was one of the bright young men who spent time at Bohr's Copenhagen Institute. His September 1925 paper presented a mathematical method to calculate these probabilities at various locations. Heisenberg structured this as a matrix, essentially a grid of probabilities. It was tedious, but it worked so well that Heisenberg won the Nobel Prize in 1932 "for the creation of quantum mechanics."

Hot on Heisenberg's heels was Erwin Schrödinger.[128] He transformed Heisenberg's description of quantum mechanics into a wave-oriented equation rather than Heisenberg's matrix-oriented equation. Schrödinger's equation, published in 1926, was much easier to work with. But still, Schrödinger's equation had some holes in it. For one thing, it didn't account for special relativity.

Paul Dirac[129] was a nerd's nerd. He didn't talk much, but when he did choose to speak, every word was meaningful. Fellow students at Cambridge proposed a unit of measurement that they called "a Dirac" which consisted of uttering one word per hour. It has been suggested that in *our* day and age, he would have been labeled as autistic. And brilliant. He would have been labeled as brilliant in any age.

Dirac fused *special* relativity with Schrödinger's equation to produce what is known as Quantum ElectroDynamics (QED), or simply as "the Dirac equation." Whereas the earlier quantum equations only accounted for quantum particles of force (like photons), Dirac's equation brought quantum particles of matter into the fold. Einstein called it *"the most logically perfect presentation of quantum mechanics."* For somebody as nerdy as Dirac, you can't beat praise like that.

Dirac's formulation of quantum mechanics was the first glimmer of what is called quantum field theory (QFT). In QFT, particles are associated with "quantum fields." There is one quantum field for each type of elementary particle. These quantum fields are similar to a magnetic field or a gravitational field. Like gravity or magnetism, these quantum fields extend throughout spacetime. Each type of quantum field hosts innumerable quantum particles. Each quantum particle is a local excited state in its quantum field. A photon is a local excited state in the quantum electromagnetic field. An electron is a local excited state in the quantum electron field. Each quantum field is boundless and universal. Each quantum particle is a local excitation in its associated quantum field. This "local excitation" is what is described by the particle's wave function. The local excitation *is* the wave.

To me, the idea that a quantum particle travels like a wave dovetails nicely with the idea that the quantum particle itself is a local excitation in its associated field. A quantum particle's wave function is the description of how that particle travels like a wave across its quantum field of existence.

The Collapse of the Wave Function

But how does a photon in the double-slit experiment go from traveling like a wave, to interacting like a particle when it hits the detector?

This question haunts physicists to this day. Some call this transformation from wave behavior to particle behavior **wave function collapse**. Others call it **the measurement problem.**

One way of looking at it is to say that the photon truly exists as its wave function (all spread out) while it is traveling. Then when it interacts with something, it instantaneously transforms itself to a single point in spacetime. This is **the collapse of the wave function** school of thought.

Another way of looking at it is to say that the photon is always a particle, and its wave function is only a guide. The particle rides its wave function like a surfer rides an ocean wave. This is the **pilot wave** theory.

For an excellent, but somewhat nerdy, video on the range of scientific ideas on how to interpret the collapse of the wave function, I highly recommend the PBS Space Time episode at this link.[130]

Antimatter: Predicted and Found

Dirac's quantum field theory also had holes in it, but Dirac's holes served a purpose. Dirac's equation, published in 1928, contained a "theory of holes" which predicted antimatter. It predicted that for every type of particle possessing an electrical charge, there is a corresponding type of particle with all the same attributes, except that its electric charge is reversed. So a proton (having a positive electrical charge) has a corresponding antimatter twin, the antiproton, which has a negative electrical charge. An electron (which has a negative electrical charge) has an antimatter twin with a positive electrical charge. For historical reasons, this antimatter electron can be called either a positron or an antielectron. Positron, antielectron, same thing.

The kicker is that if a particle and its antiparticle (electron and positron, for instance) ever interact, all of the energy bound up in their combined masses will be released. It's that $\mathbf{E=mc^2}$ thing again. KABLOOIE!

"Antimatter" sounds like a seriously absurd idea to me. But in 1932 an American physicist named Carl Anderson detected positrons in the cascade of particles left by cosmic rays. Antimatter particles have been regularly observed in particle accelerators for almost a century. Antimatter has also been put to use in the PET scans that are done in hospitals. PET stands for Positron Emission Tomography. PET scans can reveal the condition of the heart, the condition of the brain, and detect cancer tumors. So I guess that I have to accept nature as She is: absurd.

Standing Waves And Atomic Orbitals

Earlier in this appendix we started looking at the internal structure of an atom. We saw how it was impossible for an atom's electrons to orbit the nucleus in the same way that the Moon orbits the Earth. Niels Bohr patched up that problem with an atomic model requiring electrons to inhabit only certain *quantum* orbits.

Now that we have the wave nature of quantum particles under our belt, we are ready to take the final step to a fully functioning model of the atom. Erwin Schrödinger took this next step in 1926. He incorporated the *wave* nature of the electron into the model. This model made the unique prediction that hydrogen atoms subjected to an external magnetic field would have their fingerprint in the rainbow split up. Where a single line in the spectrum would normally occur, hydrogen atoms in a magnetic field would show multiple lines, close together. This matched earlier experimental results. No other atomic model could explain this effect.

This wave-based model of the electron's orbit does not look like an "orbit" at all. Physicists coined several other phrases, just so they wouldn't have to say "orbit" when talking about how an electron is held around an atom's nucleus. So if you see the phrases **atomic orbitals**, **orbital shells**, **electron shells**, or just plain **orbitals** — that's this wave-based model of how an electron is held around an atom's nucleus.

To understand these atomic orbitals, let's start by looking at **standing waves**. Standing waves are a human-scale counterpart for the wave-based quantum model of the atom.

If you've ever looked at a plucked guitar string, you've looked at a standing wave.

When an open guitar string is plucked it vibrates in a wavelength that is the full length of the string. This standing wave makes the lowest note that the string can create. It's called the string's 1st harmonic.[131]

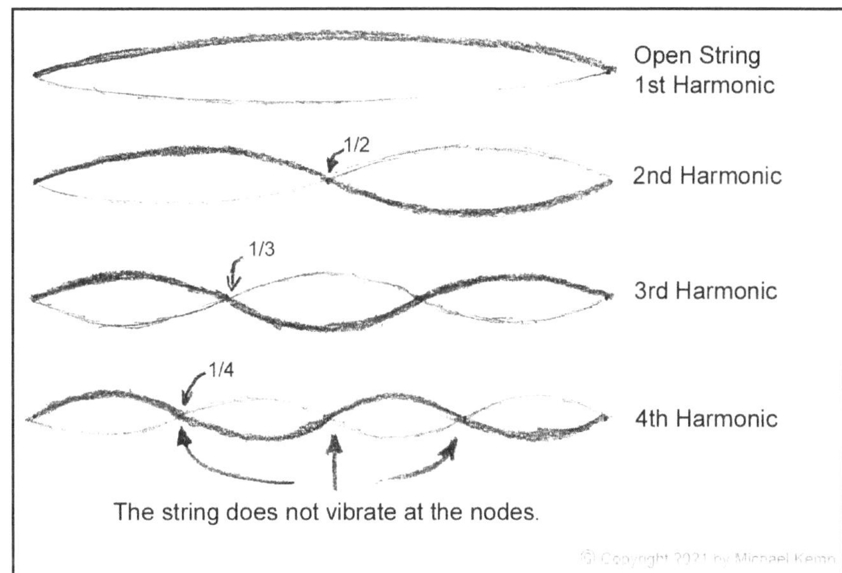

To produce the 2nd harmonic: you lightly touch the string at ½ its length with a finger of one hand, then pluck the string with a finger of the other hand *at the same time that you release your light touch*. This will cause the string to vibrate at half the wavelength and twice the frequency of the 1st harmonic.

The other harmonics can be produced in a similar manner.

I'll draw your attention to two things:

1. Harmonics (the standing waves of the guitar string) only occur at simple fractions of the string (whole, ½, ⅓, ¼ …). This is similar to how atomic orbitals only occur at distinct quantum energy levels (1, 2, 3, 4…). The harmonics, like the quantum energy levels, do not occur at just any old value.

2. In the illustration above, I have noted that the higher harmonics have "nodes" in the vibration. These are areas of the vibrating guitar string that do *not* vibrate. For instance, the entire string vibrating at the 3rd harmonic includes two nodes that do not vibrate.

You can think of the atomic orbital as a standing wave. But true to the nature of quantum mechanics, atomic orbitals are bizarre.[132]

This illustration shows only six of the hydrogen atom's many possible orbitals. Each of these six drawings is one single orbital. These orbitals each represent a different energy level. The hydrogen atom has many more orbitals than I am showing here. This is just a sample.

Take the bottom right orbital for example, the one who's center looks like a four leaf clover. The places where a photon would be likely to interact with the hydrogen atom's single electron in this orbital are dark red. Places that are white are places where a photon would not be likely to interact with the electron. The white areas are like the nodes in the vibrating guitar string. The white spaces and white lines separate the dark red areas into islands, but all of the red areas are one single orbital. An electron in this "clover leaf" orbital can be found in any of those dark red islands. It is not confined to a single island. *The electron is in the entire orbital.*[133]

It's obvious that the circular orbits shown in classic illustrations of atoms are a gross oversimplification. An atom's electrons exist as a wave function, a standing wave, around the nucleus. They are not little satellites going in circles.

A Few of Hydrogen's Atomic Orbitals

The Quantum Wave Version

Each of the six orbitals shown is a cross-section of the full 3D shape of the orbital. The atomic orbital that hydrogen's single electron inhabits depends on the energy level of the electron. A single atomic orbital can have separate high amplitude areas, shown in dark red, where there is a high probability of interacting with the electron. Areas that are white indicate places where you are unlikely to interact with the electron.

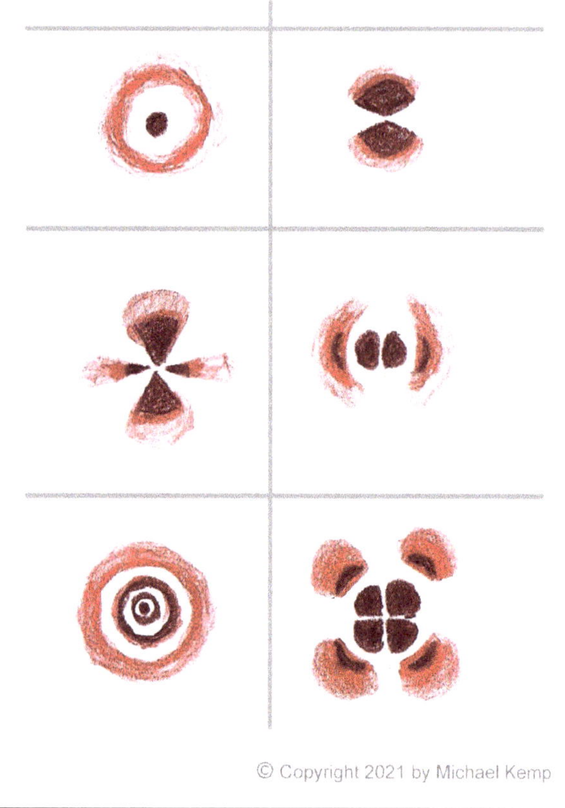

But is Quantum Mechanics Real?

Quantum mechanics eventually developed to the point that it described all of the atoms that were known at the time. It also describes all of the atoms that have since been discovered.

Quantum mechanics described all of the elementary particles that were known, and all of the elementary particles that were later discovered.

While quantum mechanics was developed in the early 1900s, the final elementary particle that it predicted was not confirmed until 2012. It took that long because revealing the Higgs particle required energy levels only attained by the Large Hadron Collider.[134] Even then, it required specialized equipment and a couple of years of data collection to be sure that the results were solid.

A Standard Model of All the Bits

The experimental discovery of the electron in 1897, and the gradual acceptance in the following decades that light was made of photons, were the first hints that matter and force come in quantum particles.

The electron showed itself to be a quantum particle of matter. The photon showed itself to be the quantum particle of electromagnetic force.

The atomic nucleus proved to be constructed of protons and neutrons. Protons and neutrons were later revealed to be composed of quarks (quantum particles of matter) held together by gluons (quantum particles of the strong force).

Quantum mechanics predicted all of these quantum particles, and more. The result is a collection of the truly elementary quantum particles called **The Standard Model of Particle Physics**.[135]

Contemplating all of these elementary particles, and all of the composite particles that can be constructed from them, Robert Oppenheimer jokingly called it "the subnuclear zoo." To this day, this vast array of elementary particles and the particles constructed out of them (such as protons and neutrons) is referred to as "the particle zoo."

Next is an illustration of The Standard Model of Particle Physics.

Note that I did not include the antimatter versions of particles.

It's also worth noting that I only listed the gluon once, while in reality there are eight "color charge" versions of gluons.

The Standard Model of Particle Physics is quite complex enough without throwing all the gluon colors and all of the antiparticles in!

The Standard Model of Particle Physics

Quantum Mechanics produced this (experimentally verified) list of the truly elementary particles of matter and force in our universe. Each particle can be thought of as a local excitation in a universal field of existence for that particular type of matter or force. For instance, the Higgs particle is a local excitation in the Higgs Field. The Electron is a local excitation in the Electron Field. The Up Quark is a local excitation in the Up Quark Field. And so forth and so on, each particle being an excitation in its own field of existence — a field of existence that extends all the way across our universe.

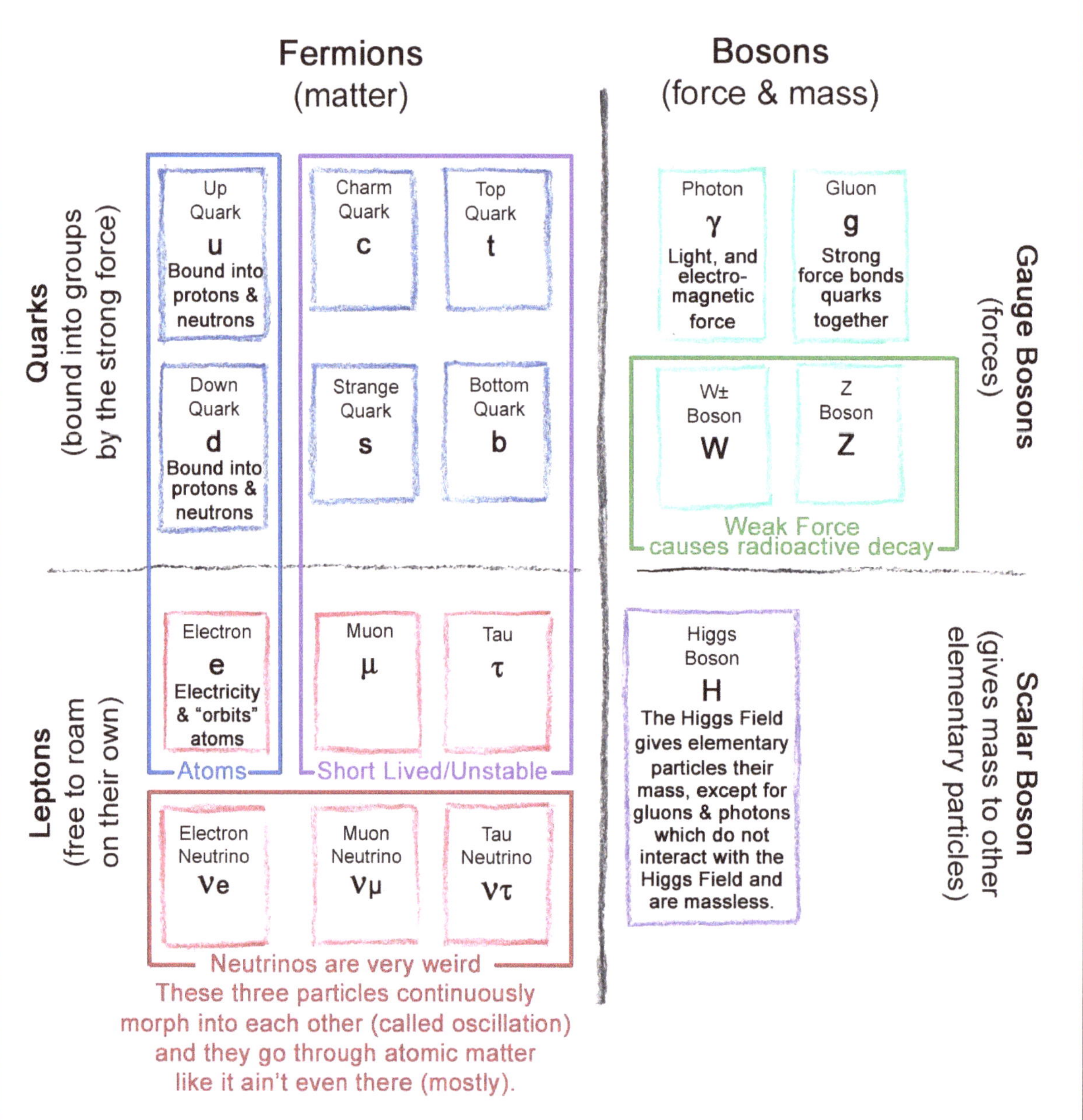

The Standard Model of Particle Physics reached its current form in the 1970s. These truly uncuttable quantum particles fall into two main groups. Elementary particles of *matter*, called **fermions** (which Dirac named in honor of physicist Enrico Fermi). And elementary particles of force, called **bosons** (which Dirac named in honor of physicist Satyendra Nath Bose).

Fermions (matter) are further divided into two categories: **quarks** and **leptons**.

Quarks interact with gluons (gluons are a particle of force, a boson). It gets complicated, but basically, gluons bind three quarks into a proton or into a neutron. Whether the three quarks form a proton or neutron depends on the type of quarks involved. This binding by gluons is called the **strong force**.

The interaction of gluons with the quarks inside protons and neutrons sometimes creates free quark pairs that are exchanged between protons and neutrons. This exchange is responsible for gluing protons and neutrons together in the atom's nucleus. This is called the **strong *nuclear* force**.

Protons have a positive electrical charge, and neutrons are electrically neutral. Protons give the nucleus a positive electrical charge.

The other type of matter, **leptons**, does not interact with gluons. The most famous lepton is the electron. Opposite electrical charges attract each other, and an electron, having a negative electrical charge, is attracted to the positively charged atomic nucleus.

The short-ranged strong nuclear force keeps the protons and neutrons of the nucleus bound tightly together. The much weaker (but long-range) electromagnetic force attracts the electron to the much more massive atomic nucleus.

Moving right along, the other major type of elementary particles, the **bosons** (forces), are also divided into two categories, **gauge** and **scalar**.

The **gauge** bosons carry three of the forces of nature: **electromagnetism** (transmitted by photons), the **strong force** (transmitted by gluons), and the **weak force** (transmitted by the W and Z bosons). The weak force is responsible for the radioactive decay of atoms. The fourth force of nature, **gravity**, is not accounted for by the Standard Model of Particle Physics.

The second category of bosons (the **scalar** boson) has only one member: the Higgs boson. The Higgs quantum field, which produces the Higgs boson, gives mass to other elementary particles. The Higgs quantum field does not interact with photons and gluons, so those particles are massless.

Please note that the Higgs field is not the sole, or even primary, contributor to the mass of composite particles. Composite particles like protons, neutrons, and atoms derive most of their mass from the energy that is required to keep all their parts bound together. That binding energy is most of their mass. You can use basic algebra to solve **$E=mc^2$** for mass (**m**) and get **$m=E/c^2$** which translates in non-nerd speak to "if you bind up a whole lot of energy it shows up as a little bit of mass."

So most of the mass of protons, atoms, apples, you, me, Earth, the Sun, ice cream, etc., comes from the energy bound up in keeping all their particles glued together.

But yeah, the Higgs field also adds a little mass to the particles that it interacts with.

All of the various elementary particles predicted in the Standard Model have been experimentally verified. The observation of the Higgs boson in 2012 confirmed the final piece of the Standard Model of Particle Physics.

It is quite possible that other particles will be predicted and/or discovered by experiment as time goes by. In fact, other elementary particles are already predicted by a hypothesis called Supersymmetry. However, these supersymmetric particles have not yet been experimentally observed.

One missing piece of the Standard Model of Particle Physics is that it has no boson to transmit the force of gravity. The Standard Model of Particle Physics leaves gravity to general relativity's bending of spacetime. I should note that string theory modifies quantum mechanics in a way that adds such a particle, the graviton.

At any rate, the Standard Model of Particle Physics is the *experimentally confirmed* list of the most basic, *truly uncuttable* bits of matter and force from which everything in our universe is built!

The Uncertain Wavicle (Heisenberg's Uncertainty Principle)

Heisenberg's uncertainty principle[136] is one of the most well-known, and to my mind most misunderstood, consequences of quantum mechanics.

The uncertainty principle states that there are certain pairs of *attributes* of a quantum particle that are tied to each other in such a way that the more tightly one of them is constrained, the more uncertain the other attribute is. In nerd-speak these pairs of attributes are called "complementary" or "conjugate" variables.

The classic example of a pair of complementary variables is the position and the momentum of a quantum particle. So the more precisely a quantum particle's position is constrained, the more uncertain is its momentum. And it works the other way around too. The more precisely a quantum particle's momentum is constrained, the more uncertain is its position.

That sounds absurd from the perspective of common sense, but we are talking about *very* small amounts of uncertainty. This is an effect that only shows up in the quantum realm.

One version of the formula for the uncertainty principle is written as $\boldsymbol{\sigma_x \sigma_p \geq \hbar/2}$. The $\boldsymbol{\sigma_x}$ means "how constrained is the particle's position" and the $\boldsymbol{\sigma_p}$ means "how constrained is the particle's momentum." The $\geq \boldsymbol{\hbar/2}$ part means that the precise value of *the combination* of the position and momentum is just slightly uncertain. That value of uncertainty ($\boldsymbol{\hbar/2}$) is so small that, in the human scale of existence, it might as well be zero. But in the quantum realm, it is significant.

At the rock-bottom root of reality, in the wave function world of atoms and quantum particles, there is no absolute certainty. This is due to the quantum particle's wave nature.

It should be no surprise that a lot of people find this version of the uncertainty principle unacceptable. Even Heisenberg himself did not like it. Some folks have presented the uncertainty principle as meaning that you cannot *observe* such a small particle without the observation itself disturbing the particle's position and/or momentum. For instance, to observe an atom with an electron microscope, the microscope bounces electrons off of the atom. The impact of those electrons on the atom being studied will alter the atom's position and momentum.

In this interpretation, the uncertainty principle can be re-framed as $\boldsymbol{\Delta x \Delta p \geq \hbar/2}$. The $\boldsymbol{\Delta}$ means "change in…", so this version reads "the change in position multiplied by the change in momentum is always greater than or equal to half of the reduced Planck's constant."

Personally, I prefer the wave function version of the uncertainty principle, $\boldsymbol{\sigma_x \sigma_p \geq \hbar/2}$. To me, this dovetails nicely with the experimental evidence that quantum particles travel as a wave until they interact with their surroundings. To my mind, the quantum realm is uncertain, not because it is difficult to measure quantum particles (which it is), not because the act of measuring disturbs quantum particles (which it does), but because uncertainty is part of a quantum particle's true nature. The position and momentum of a quantum particle are uncertain, not because it is being observed, but because a quantum particle exists (most of the time) as a wave function, as a local excitation in its quantum field of existence.

Other Quantum Weirdness

There are a few other aspects of quantum mechanics that I'll just mention in passing: virtual particles, entanglement, and the hypothesis that the fabric of spacetime is held together by wormholes between entangled virtual particles.

Virtually Invisible

First off, what's a "virtual" particle? What we call a "real" particle is a local excitation in that particle's quantum field of existence. For instance, a real photon is a local excitation in the electromagnetic quantum field. A *virtual* photon is pretty much the same, except that a virtual photon exists on such a short timescale that its existence hides behind the veil of Heisenberg's uncertainty principle. A virtual particle is here-and-gone almost instantaneously, falling under the $\hbar/2$ of uncertainty's $\sigma_x \sigma_p \geq \hbar/2$ limit.

One place where virtual particles appear is in how real particles interact with each other. For example, when two negatively charged electrons repel each other, that electromagnetic repulsion happens by exchanging one or more *virtual* photons between the two electrons. This exchange supplies the electromagnetic force that repels the two electrons from each other.

Another example is that to keep three quarks bound together to form a proton, virtual gluons are continuously exchanged between the quarks. Yes, *virtual* gluons glue the quarks together so that they form a stable proton.

Another place where you might hear about virtual particles is, well, everywhere. According to the uncertainty principle, the quantum fields of existence are never completely still. Even where there is not enough local excitation in a quantum field to create a "real" particle, the uncertainty principle says that pairs of virtual particles wink in and out of existence throughout the field, continuously.

Virtual particles are impossible to detect directly, but they are our best explanation for a number of observed phenomena, such as the behavior of subatomic particles in experiments at particle accelerators.

Entangled Relationships

One consequence of the waveform nature of quantum particles is that when two particles' waveforms overlap, they can form a lasting relationship called entanglement. Their quantum attributes become shared. This entanglement continues even when the particles are separated by some distance. According to entanglement, if you effect the state of one entangled particle, the other entangled particle is instantly effected as well. This instant communication, of course, violates spacetime's speed of causality.

Einstein, having gone all-in on special relativity's maximum speed of causality, was not a fan of the concept of entanglement. But if experiment shows that a proposed idea is real, that beats whatever it is that you *want* to be real.

Entanglement has been verified by experimental testing. Here's a video from Fermilab that covers entanglement, Einstein's alternative explanation, and what actual experiments have shown.[137]

Wormholes in Quantum Space

One proposed explanation for the instantaneous nature of quantum entanglement is a mind-bending mashup between general relativity and the quantum realm.

In exploring the possibilities of warped spacetime, Einstein and a fellow physicist Nathan Rosen proposed that two distant points in space could be pinched together in such a way as to create a bridge between them. This would eliminate the distance that originally separated those two points. This is called a wormhole.

It has been suggested that entangled particles are actually connected by quantum-sized wormholes. This would explain their instantaneous communication. Through the wormhole, there would be no distance between them.

Remember those pairs of virtual particles that wink in and out of existence in their quantum fields? The fact that they are created as a pair means that their waveforms are entangled.

One of the more mind-blowing suggestions is that these quantum fields of here-and-gone virtual particles, with their connecting wormholes, are the fabric that holds space together. There are mathematical arguments in favor of this hypothesis, but I have no idea how you could test it. Someone would have to come up with a unique prediction that our current technology could test, before this could move from being a hypothesis to being a viable theory.

That's Quite Enough, Thank You

And with that final sampling of quantum absurdities, I'll wrap up these appendixes on quantum mechanics.

Nothing about quantum mechanics seems quite sensible. Quantum mechanics was forced upon us by the facts, by experimental results. While common sense might rebel, observed phenomena required new thinking about how nature works.

Nature doesn't really care what our opinion is. If we want to know how nature works, it's our task to figure out what nature does, not what we wish it did. If you want to get real, you'll just have to deal.

I'll be honest. There are two reasons that I've bothered you with quantum mechanics in this book about gravity. First, since quantum mechanics (supported by extensive experimental evidence) does not play well with general relativity (also supported by extensive experimental evidence), we may be missing something about gravity. Einstein's gravity might not be the final answer. Second, quantum mechanics is just too crazy amazing to leave out. I had to share the story of how it was created, and how its outrageous and unique predictions were confirmed by experiment after experiment.

I hope you've enjoyed reading the story as much as I've enjoyed writing it!

Full disclosure: The results of quantum mechanical experiments (the observed phenomena) are clear, repeatable, and verified. But the *interpretation* of quantum mechanics splits into various camps.

The three main camps are the Copenhagen Interpretation, the Pilot Wave school of thought, and the Many Worlds Interpretation. I lean toward the Copenhagen Interpretation (mostly), so what I have presented reflects that way of thinking about quantum mechanics. You might be intrigued by the Many Worlds or the Pilot Wave way of looking at things. Many Worlds proposes that every possible option that a particle has is realized in an alternate universe. This would spawn an unimaginable number of universes in every instant of time. Pilot Wave interpretation, on the other hand, keeps the idea of a single knowable universe, with particles as tiny points of matter, guided by an associated wave function. Here's a link to PBS Space Time's Pilot Wave explainer.[138]

Review

This appendix begins with Thomas Young's double-slit experiment. This experiment demonstrated that photons of light travel all spread out as waves and even interfere with themselves after passing through the double slits. And yet each photon impacts the detector screen at a specific location, as a particle.

Anders Ångström discovered that when pure hydrogen gas is excited until it produces light, the spectrum of that light spikes at a few select wavelengths. To turn the tables, when a full spectrum of light is projected through cold hydrogen gas, the gas absorbs light at those same wavelengths. This is the "spectrum" of hydrogen. This is the hydrogen atom's fingerprint in the rainbow. It's not just hydrogen. Any pure substance, any specific type of atom, has its own unique fingerprint in the rainbow.

Following up on a "that's funny" moment in the experiments Heinrich Hertz had made decades earlier, Albert Einstein revealed that electrons are bound to their atoms at quantized energy levels. The electrons in Hertz's metal antenna *only* responded to photons having the proper amount of energy to free them from their bond.

Ernest Rutherford proposed the first model of the atom that included a massive central nucleus with a positive electrical charge, surrounded by light-weight orbiting electrons. He based this idea on the results of experiments, analyzing how beams of alpha particles were scattered off target atoms. This is the image of an atom that is still used extensively today, even though it was shown to be flawed.

Niels Bohr recognized that Rutherford's model would not produce stable atoms. Bohr took Planck and Einstein's description of light and electron bonds as being quantized, and used that idea to create a new model of the atom. In Bohr's model, the electrons are restricted to quantized **atomic orbitals**. His equations for these quantized atomic orbitals not only produced a model for stable atoms, it explained Ångström's discovery of the hydrogen atom's fingerprint in the rainbow. An electron in one of the atom's orbitals can only jump to a higher energy orbital by absorbing and destroying a photon of the exact energy difference between those orbitals. An electron in a higher-energy orbital can only jump to a lower-energy orbital by producing and releasing a photon of the exact energy difference between those orbitals.

And then there's the *wave* nature of quantum particles of matter and force.

It was established in 1801 that light travels like a wave. It was also shown (in the 1920s) that an electron also travels like a wave. Both quantum particles of force (like photons) and quantum particles of matter (like electrons) travel all spread out, like a wave. And yet when they interact with the world around them, they interact at a specific location, like a particle. This is the wave-particle duality, and why quantum particles are sometimes called wavicles.

Quantum mechanics, which was created to describe the behavior of these smallest bits of existence, developed into a theory of fields and waves. Each *type* of quantum particle has an associated quantum field of existence. Each individual quantum particle is a local excitation in their respective quantum field. Each quantum particle is also described by a wave function which describes the attributes of that particular particle, such as their position and momentum. This quantum wave function does not give specific values, but rather it gives the *probability* of interacting with that particle in various states. For instance, the *probability* of interacting with that quantum particle at various positions with various momentums.

When the wave nature of quantum particles is applied to the electrons that are bound to the nucleus of an atom, it transforms the atomic orbitals into clouds of probability. An individual electron in an individual orbital is effectively spread out across the entire orbital, even when a single orbital is divided up into separate islands of probability.

But is quantum mechanics real?

As quantum mechanics was refined, it predicted antimatter particles — which were subsequently found in nature and (decades later) in particle accelerators.

Quantum mechanics also described all of the elementary, uncuttable, quantum particles that were known at the time, and all that were later discovered.

That combination of unique prediction and experimental confirmation is the scientific method's gold standard for calling your description of nature a valid *theory*.

One of the more unsettling predictions of quantum mechanics is Heisenberg's uncertainty principle. Based on the wave function of quantum particles, the uncertainty principle states that there are pairs of attributes that cannot be both precisely defined at the same time. The more precisely one of the pair is defined, the more freedom the other attribute has to be less defined. The pair of quantum attributes that is usually cited is a quantum particle's position and its momentum. Folks often pass this off as an uncertainty in *measuring* such tiny bits of matter and force, but the uncertainty principle is based on the quantum wave function. It does not rely on the vagaries of external measurement.

For a visual simulation of how the quantum wave function evolves over time, how pairs of attributes such as position and momentum are tied to each other, and ultimately how the uncertainty principle works, follow this link.[139] A couple of heads-ups about this video:

- It moves along quickly. Be prepared to pause, rewind, and replay sections.
- For the majority of the video they show position versus *velocity* rather than position versus *momentum*. Since momentum is just mass times velocity, velocity is a decent placeholder for momentum. They come back to momentum at the end of the video.

That being said, it's a great visual presentation of the wave function.

The concept of virtual particles that pop in and out of existence is pretty strange, but it's a handy way to model how real particles interact. And it explains the collision results recorded in particle accelorators.

Entanglement is a very cool feature of quantum mechanics. But don't get carried away. It only functions in the quantum realm.

Put entangled virtual particle pairs together with quantum wormholes, and you just might have the fabric of space. But that's just speculation at this point.

And yes, the clash between general relativity and quantum mechanics means that Einstein's gravity might not be the last word on *why stuff falls down*.

Activity

The Double-Slit Experiment

Yes, you too can reveal the wave nature of light in the comfort of your own home! The details are a little touchy, so you may want to watch my video of this activity before you attempt it yourself.

Supplies for this activity:

- A laser pointer, the one you use to play with the cat will work just fine. A red laser pointer works best, but other colors can work too.
- A piece of cardboard (or liner board) that you can cut a hole in.
- A piece of white paper to project the laser light onto.
- Electrician's tape, because electrician's tape is opaque and the edge of the tape is smooth and straight.
- A thin piece of wire. I've had success with stripping the wire out of a twist-tie that came with a box of plastic bags. Some twist-ties have a wire in them. Scrape the paper off of the wire and clean the wire up so that it is smooth and straight.
- Scissors to cut the tape.
- And a dime would come in handy.

To create the double-slit experiment:

1. Cut a small hole in the cardboard, ½" square works great.
2. Cut two pieces of electrician's tape, let's say 2" long each.
3. Place the two pieces of electrician's tape parallel to each other, over each side of the hole in the cardboard. Leave a space between them that is the width of a dime, or a little less than 1/16". This creates a single slit for light to pass through.
4. Use a thin piece of wire. I've successfully used the cleaned-up wire from a twist-tie. The strand of wire should be as clean and close to straight as you can make it.
5. Cut two more pieces of electrician's tape, about an inch long.
6. Use these one-inch pieces of tape to secure your piece of wire down the center of the slit that you made with the first two pieces of tape. You should see a tiny slit along *each side* of the wire. You have now created the famous double-slit.
7. Place the piece of white paper on a table. Turn off any bright lights so that the paper is in shadow. Hold your cardboard double-slit a couple of feet above the paper. Shine your laser pointer through your double-slit and onto the paper.

8. If all has gone well, once you fiddle with where the laser is hitting the double-slit, you will see a series of short vertical lines. If not, please review my video of this activity and look for any differences between my setup and yours.

If light traveled like a particle or as a tiny ray — in straight lines — then you would expect to see a single blob of light on the paper or maybe two blobs, one for each slit. The fact that you see a series of short vertical lines shows that an interference pattern has been created. This is a physical demonstration that light travels as a wave.

Here's the video of my recreation of the famous double slit experiment:
https://www.whatacuriousworld.com/appendix-b/#VideoB

Links

Deeper Dive
119 "Corpuscular theory of light - Wikipedia."
https://en.wikipedia.org/wiki/Corpuscular_theory_of_light.

Deeper Dive
120 "Light - Young's double-slit experiment | Britannica."
https://www.britannica.com/science/light/Youngs-double-slit-experiment.

Biography
121 "Anders Jonas Ångström - Wikipedia."
https://en.wikipedia.org/wiki/Anders_Jonas_%C3%85ngstr%C3%B6m.

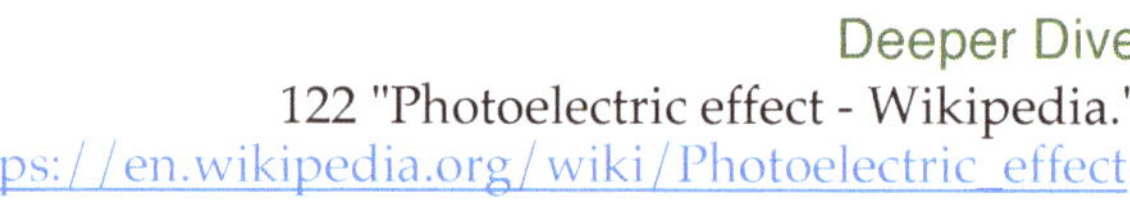

Deeper Dive
122 "Photoelectric effect - Wikipedia."
https://en.wikipedia.org/wiki/Photoelectric_effect.

Great Video
123 "Photoelectric effect Demonstration"
https://www.youtube.com/watch?v=v-1zjdUTu0o.

Biography
124 "Niels Bohr | Biography, Education, Accomplishments, & Facts."
https://www.britannica.com/biography/Niels-Bohr.

Deeper Dive
125. "Bohr model - Wikipedia."
https://en.wikipedia.org/wiki/Bohr_model.

Physics-Speak

126 "Davisson–Germer experiment - Wikipedia." https://en.wikipedia.org/wiki/Davisson%E2%80%93Germer_experiment

Biography

127 "Werner Heisenberg | Biography, Nobel Prize, & Facts | Britannica." 28 Jan. 2022, https://www.britannica.com/biography/Werner-Heisenberg.

Biography

128 "Erwin Schrödinger - Wikipedia." https://en.wikipedia.org/wiki/Erwin_Schr%C3%B6dinger.

Biography

129 "Paul A.M. Dirac – Biographical - NobelPrize.org." https://www.nobelprize.org/prizes/physics/1933/dirac/biographical/.

Great Video

130 "Is The Wave Function The Building Block of Reality?" https://www.youtube.com/watch?v=FP6iyVJ70OU

Deeper Dive

131 "Harmonic - Wikipedia." https://en.wikipedia.org/wiki/Harmonic

Great Video

132 "Understanding Quantum Mechanics #7: Atomic Energy Levels." 7 Nov. 2020 https://www.youtube.com/watch?v=LBTNKzZLo-s

Deeper Dive

133 "Atomic orbital - Wikipedia." https://en.wikipedia.org/wiki/Atomic_orbital

Deeper Dive

134 "The Large Hadron Collider" https://home.cern/science/accelerators/large-hadron-collider

Deeper Dive

135 "Standard Model - Wikipedia." https://en.wikipedia.org/wiki/Standard_Model

Deeper Dive

136 "Uncertainty principle - Wikipedia."

https://en.wikipedia.org/wiki/Uncertainty_principle

Great Video/Physics-Speak
137 "Quantum Entanglement: Spooky Action at a Distance"
https://www.youtube.com/watch?v=JFozGfxmi8A

Great Video
138 "Pilot Wave Theory and Quantum Realism | Space Time"
https://www.youtube.com/watch?v=RlXdsyctD50

Great Video/Deeper Dive
139 "Visualization of Quantum Physics (Quantum Mechanics)."
https://www.youtube.com/watch?v=p7bzE1E5PMY

About the Author

Michael Kemp (that would be me) was born, raised, and still resides in western Oregon. There were years spent in other states, and some minor adventures abroad, but I always came back home to Oregon.

My love for physics was kindled, over half a century ago, by my high school physics teacher: Mr. Hawken. That led me to take physics classes in college, but my professional career was in computers.

Throughout my working years, my fascination with science and physics would not leave me alone, leading me to keep tabs on science news and trends. I've found it frustrating that "news" sources often put a misleading spin on science topics.

Once I retired, my wife and I got some long-overdue camping and national parks visits checked off the to-do list. But even in blissful retirement, the gap between how science actually works, and what regular folks think of science, kept nagging at me. By telling the story of science and gravity, I've tried to make science and gravity understandable.

I hope you enjoy my little book. And I hope it gets you thinking about our vast, awesome, and fantastical universe.

Keep Well!

Michael Kemp

Website
https://www.whatacuriousworld.com/

Facebook
https://www.facebook.com/Michael.Kemp.Author

YouTube
https://www.youtube.com/@whatacuriousworld-com

Instagram
https://www.instagram.com/what_a_curious_world/

www.ingramcontent.com/pod-product-compliance
Lightning Source LLC
LaVergne TN
LVHW070933160826
845679LV00021B/1797

* 9 7 9 8 9 8 9 8 4 6 0 0 9 *